SpringerBriefs in Astronomy

Series Editors

Martin Radcliffe
Wolfgang Hillebrandt

For further volumes:
http://www.springer.com/series/10090

Karen L. Aplin

Electrifying Atmospheres: Charging, Ionisation and Lightning in the Solar System and Beyond

 Springer

Karen L. Aplin
Department of Physics
University of Oxford
Oxford
UK

ISSN 2191-9100   ISSN 2191-9119 (electronic)
ISBN 978-94-007-6632-7   ISBN 978-94-007-6633-4 (eBook)
DOI 10.1007/978-94-007-6633-4
Springer Dordrecht Heidelberg New York London

Library of Congress Control Number: 2013934386

Preface

Electrical processes take place in all planetary atmospheres. There is evidence for lightning on Venus, Jupiter, Saturn, Uranus and Neptune, it is possible on Mars and Titan, and cosmic rays ionise every atmosphere, leading to charged droplets and particles. Controversy surrounds the role of atmospheric electricity in physical climate processes on Earth; here, a comparative approach is employed to review the role of electrification in the atmospheres of other planets and their moons. This paper reviews the theory, and, where available, measurements, of planetary atmospheric electricity, taken to include ion production and ion–aerosol interactions. The conditions necessary for a global atmospheric electric circuit similar to Earth's, and the likelihood of meeting these conditions in other planetary atmospheres, are briefly discussed. Atmospheric electrification is more important at planets receiving little solar radiation, increasing the relative significance of electrical forces. Nucleation onto atmospheric ions has been predicted to affect the evolution and lifetime of haze layers on Titan, Neptune and Triton. For planets closer to Earth, heating from solar radiation dominates atmospheric circulations. Mars may have a global circuit analogous to the terrestrial model, but based on electrical discharges from dust storms, and Titan may have a similar global circuit, based on transfer of charged raindrops. There is an increasing need for direct measurements of planetary atmospheric electrification, in particular on Mars, to assess the risk for future space missions. Theoretical understanding could be increased by cross-disciplinary work to modify and update models and parameterisations initially developed for a specific atmosphere, to make them more broadly applicable to other planetary atmospheres. Electrical processes in the atmospheres of exoplanets are also discussed. This book is an updated and expanded version of a review paper published in 2006,[1] which is cited where it contains additional material.

---

[1] K.L. Aplin, Atmospheric electrification in the Solar System, *Surveys in Geophysics*, **27**, 1, 63-108 (2006). doi: 10.1007/s10712-005-0642-9

Acknowledgments

I thank Prof. R. G. Harrison (University of Reading) and Dr. S. Aigrain (University of Oxford) for helpful comments. Dr. K. A. Nicoll (University of Reading) assisted with figure production.

Contents

Chapter 1
Introduction and Scope

Abstract Comparative planetology uses the science of Earth's environment to understand other planets, and planetary observations can also be used to broaden understanding of the terrestrial environment. This introductory chapter motivates the comparative approach and introduces the relevant physical concepts.

Comparative planetary science assumes that the science of Earth's environment can be used to understand other planetary environments; similarly, observations of other planets can be used to broaden understanding of the terrestrial environment. Several aspects motivate a modern comparative study of planetary atmospheric electricity. Firstly, electrification appears to have a small effect on Earth's climate. It follows from the Boltzmann distribution that most aerosol particles in an atmosphere are charged, with the charged fraction depending only on temperature (Keefe et al. 1959). The interaction between ions and aerosol particles could affect a planet's radiative balance through electrical influences on aerosol particle changes, both on Earth (Harrison and Carslaw 2003) and elsewhere (Moses et al. 1992). Secondly, studying other planetary atmospheres could contribute to understanding the origins of life, in which lightning has been implicated (Miller 1953). Thirdly, as manned space missions to Mars appear possible (Bonnet and Swings 2004), there needs to be an assessment of the potential electrostatic hazards facing future space missions. Fourthly, the number of planets discovered outside our Solar System with atmospheres is increasing steadily (e.g. Seager and Deming 2010), and a comparative approach can be used to deduce some of their properties, as observations remain limited. Lightning and electrical processes are expected to take place in these planetary atmospheres just as they do in the Solar System, where lightning is relatively common [occuring in 4 ± 1 of 7 planetary atmospheres (Harrison et al. 2008)].

The innermost planet, Mercury, and almost all solar system moons will not be discussed further here, as they do not have atmospheres. Specifically, atmospheric electrification comprises lightning, which is caused by convection, and non-convective electrification. Non-convective electrification requires ionisation to produce electrically charged particles, which can originate from cosmic rays, radioisotope decay, or UV radiation. Cosmic rays are everywhere, so every

K. L. Aplin, *Electrifying Atmospheres: Charging, Ionisation and Lightning in the Solar System and Beyond*, SpringerBriefs in Astronomy, DOI: 10.1007/978-94-007-6633-4_1,
© Springer Science+Business Media Dordrecht 2013

1

planetary atmosphere can be expected to contain charged particles from ionisation, causing a slight atmospheric electrical conductivity in addition to any other sources of ionisation. Through interactions with the ions and electrons produced by cosmic ray ionisation, aerosol particles add complexity to atmospheric charge exchange. The latitudinal distribution of atmospheric ionisation by cosmic rays is related to a planet's magnetic field, which modulates the incident cosmic radiation (Gringel et al. 1986). In addition, the strength of the solar wind controls the deflection of cosmic rays away from a planet, so the effectiveness of this ionisation modulation over the solar cycle is related to the planet's distance from the Sun.

C.T.R. Wilson suggested that terrestrial atmospheric electrification was sustained by the existence of a global atmospheric electric circuit (Wilson 1920), resulting from the electric current flow generated by disturbed weather and ionisation in the weakly conducting atmosphere between the surface and the ionosphere, which are better conductors. The minimum parameters required for a "Wilson" global atmospheric electric circuit are the existence of an atmosphere bounded by a conductive ionosphere and surface, with a charge generation mechanism. Aplin et al. (2008) explain that the charge generation in the alternating current (ac) part of a planet's global atmospheric circuit (mostly from lightning) can be identified by the presence of extremely low frequency (ELF) resonances in the surface-ionosphere waveguide. The existence of these "Schumann resonances" also demonstrates the presence of atmospheric charge generation and a conducting upper layer. Additionally, current flow is required to confirm the existence of a direct current (dc) global circuit. Both ELF signals and aspects of the ac global circuit are directly related to the occurrence of planetary lightning, which has been reviewed extensively (e.g. Zarka et al. 2008; Yair 2012), and will not be covered in this paper beyond its association with non-convective atmospheric electrification. Necessary conditions for a dc global atmospheric electric circuit to exist on a planet can be defined as:

1. Polar atmospheric molecules, to sustain charge.
2. Charge separation, usually generated by convection from meteorological processes, is required to form the dipole structure leading to electrostatic discharge. If there are no discharges, precipitation must carry charge to ground.
3. Evidence for conducting upper and lower layers.
4. Mobile charged particles, to provide current flow.

Assuming that all planetary atmospheres have some charge separation due to ionisation, three aspects have been selected to characterise electrical systems in planetary atmospheres.

1. If convective or other processes cause sufficient charge separation for electrical discharges, meteorological changes will cause a global modulation of the planetary electrical system.
2. Aerosols, if present, are linked to local atmospheric electrification through ion-aerosol interactions.

Table 1 Summary of electrification in Solar System planetary atmospheres. Atmospheric constituents, surface temperature and pressure are from Lewis (1997)

Planet/ Moon	Principal atmospheric constituents (% by vol)	Surface pressure (ratio to Earth), T(K)	Atmospheric electrical processes			Key Refs
			Lightning	Charged aerosol	Ion nucleation?	
Venus	CO_2 96.5, N_2 3.5	92, 700	Probably	Almost certain, not yet observed	Possibility of direct "Wilson" nucleation	Borucki et al. (1982), Michael et al. (2009), Russell et al. (2007)
Earth	N_2 78.0, O_2 20.9	1,288	Cloud-to-ground, intracloud, intercloud, sprites	Yes	No "Wilson" nucleation but probably ion mediated nucleation	Harrison and Carslaw (2003), Rycroft et al. (2012)
Mars	CO_2 95.3, N_2 2.7, Ar 1.6, O_2 0.13, CO 0.08	$5 \cdot 10^{-3}$, 210	Dust discharges are likely, but no observations	Triboelectrically charged dust	Unlikely: no supersaturation or condensable species	Farrell and Desch (2001), Michael et al. (2008)
Jupiter	H_2 89.9, He 10.2	Gas giant	Detected optically and electrically	Likely	Possible, but unlikely to be significant	Capone et al. (1979), Lewis (1997)
Saturn	H_2 96.3, He 3.3	Gas giant	Detected optically and electrically	Likely	Similar to Jupiter?	Dyudina et al. (2010)
Titan	N_2 94, CH_4 6, H_2 0.2, CO 0.01	1.5, 94	Probably not	Yes	Almost certainly, in upper atmosphere	Fulchignoni et al. (2005), Waite et al. (2007)
Uranus	H_2 82.5, He 15.2, CH_4 2.3	Ice giant	Possible; suggestive radio signals from Voyager	Likely: limited by abundance of neutral aerosols	Similar to Neptune?	Miner (1998)

(continued)

Table 1 (continued)

Planet/ Moon	Principal atmospheric constituents (% by vol)	Surface pressure (ratio to Earth), T(K)	Atmospheric electrical processes			Key Refs
			Lightning	Charged aerosol	Ion nucleation?	
Neptune	H_2 80.0 He 19.0 CH_4 1.5	Ice giant	Possible; whistlers detected by Voyager	Likely: limited by abundance of neutral aerosols	Wilson nucleation of diacetylene possible	Moses et al. (1992)
Triton	N_2 99.99 CH_4 0.01	10^{-5}, 37	Highly unlikely, too cold	Possible; haze could be charged	Wilson nucleation predicted	Delitsky et al. (1990)
Pluto	CH_4 N_2	3×10^{-6}, 50	Highly unlikely, too cold	Similar to Triton?	Similar to Triton?	Hubbard (2003)

3. If ions are implicated in the formation or removal of aerosol particles, there is a potential link between cosmic ray ionisation and the planet's atmospheric radiative balance.

This review begins with a brief discussion of atmospheric electrification at Earth, to introduce the relevant physical processes. Table 1.1 gives an overview of solar system planetary atmospheres to be discussed, and their electrical systems, in terms of the three principal aspects listed above. The book is organised moving outwards from the Sun from Venus to Pluto and concluding with a chapter on exoplanets. Measurements and theoretical predictions of ionisation and other electrical processes in each planetary atmosphere are summarised, and the likelihood of a global atmospheric electric circuit on each planet discussed. Two moons with atmospheres and probable electrical activity have been included— Saturn's moon Titan, and Neptune's largest moon, Triton. Pluto is also considered, even though it was "demoted" to a dwarf planet by the International Astronomical Union in 2006 (see http://www.iau.org/public/pluto/), due to its similar atmosphere to Triton.

References

K.L. Aplin, R.G. Harrison, M.J. Rycroft, Investigating Earth's atmospheric electricity: a role model for planetary studies. Space Sci. Rev. **137**(1–4), 11–27 (2008). doi:10.1007/s11214-008-9372-x

R.M. Bonnet, J.P. Swings, *The Aurora Programme* (ESA Publications Division, Noordwijk, 2004)

W.J. Borucki, Z. Levin, R.C. Whitten, R.G. Keesee, L.A. Capone, O.B. Toon, J. Dubach, Predicted electrical conductivity between 0 and 80 km in the Venusian atmosphere. Icarus **51**, 302–321 (1982). doi:10.1016/0019-1035(82)90086-0

L.A. Capone, J. Dubach, R.C. Whitten, S.S. Prasad, Cosmic ray ionisation of the Jovian atmosphere. Icarus **39**, 433–449 (1979). doi:10.1016/0019-1035(79)90151-9

M.L. Delitsky, R.P. Turco, M.Z. Jacobson, Nitrogen ion clusters in Triton's atmosphere. Geophys. Res. Lett. **17**(10), 1725–1728 (1990). doi:10.1029/GL017i010p01725

U.A. Dyudina, A.P. Ingersoll, S.P. Ewald et al., Detection of visible lightning on Saturn. Geophys. Res. Letts. **37**, L09205 (2010). doi:10.1029/2010GL043188

W.M. Farrell, M.D. Desch, Is there a Martian atmospheric electric circuit. J. Geophys. Res. **E4**, 7591–7595 (2001). doi:10.1029/2000JE001271

M. Fulchignoni, F. Ferri F, F. Angrilli et al., In situ measurements of the physical characteristics of Titan's environment. Nature **438**(8), 785–791 (2005)

W.J. Gringel, J.M. Rosen, D.J. Hofmann, Electrical Structure from 0 to 30 Kilometers, in *The Earth's Electrical Environment*, ed. by E.P. Krider, R.G. Roble (National Academy Press, Washington, DC, 1986)

R.G. Harrison, K.S. Carslaw, Ion-aerosol-cloud processes in the lower atmosphere. Rev. Geophys. **41**, 3 (2003). doi:10.1029/2002RG000114

R.G. Harrison et al., Planetary atmospheric electricity. Space Sci. Rev. **137**, 5–10 (2008). doi:10.1007/s11214-008-9419-z

W. Hubbard, Pluto's atmospheric surprise. Nature **424**, 137–138 (2003). doi:10.1038/424137a

D. Keefe, P.J. Nolan, T.A. Rich, Charge equilibrium in aerosols according to the Boltzmann law. Proc. Roy. Irish Acad. **60**, 27–45 (1959)

J.S. Lewis, *Physics and Chemistry of the Solar System* (Academic Press, San Diego, 1997)

M. Michael, S.N. Tripathi, W.J. Borucki, R.C. Whitten, Highly charged cloud particles in the atmosphere of Venus. J. Geophys. Res. **114**, E04008 (2009). doi:10.1029/2008JE003258

S.L. Miller, A production of amino acids under possible primitive earth conditions. Science **117**, 528–529 (1953). doi:10.1126/science.117.3046.528

E.D. Miner, *Uranus: the planet, rings and satellites*, 2nd edn. (Wiley-Praxis, Chichester, 1998)

J.I. Moses, M. Allen, Y.L. Yung, Hydrocarbon nucleation and aerosol formation in Neptune's atmosphere. Icarus **99**, 318–346 (1992). doi:10.1016/0019-1035(92)90149-2

C.T. Russell, T.L. Zhang, M. Delva et al., Lightning on Venus inferred from whistler-mode waves in the ionosphere. Nature **450**, 661–662 (2007). doi:10.1038/nature05930

M.J. Rycroft, K.A. Nicoll, K.L. Aplin, R.G. Harrison, Global electric circuit coupling between the space environment and the troposphere. J. Atmos. Sol-Terr. Phys. **90–91**, 198–211 (2012). doi:10.1016/j.jastp.2012.03.015

S. Seager, D. Deming, Exoplanet atmospheres. Ann. Rev. Astron. Astrophys. **48**, 631–672 (2010). doi:10.1146/annurev-astro-081309-130837

J.H. Waite, D.T. Young, T.E. Cravens, et al., The process of tholin formation in Titan's upper atmosphere, Science **316**, 870 (2007). doi:10.1126/science.1139727

C.T.R. Wilson, Investigation on lightning discharges and on the electric field of thunderstorms. Phil. Trans. Roy. Soc. London A **221**, 73–115 (1920)

Y. Yair, New results on planetary lightning. Adv. Space Res. **50**, 293–310 (2012). doi:10.1016/j.asr.2012.04.013

P. Zarka, W.M. Farrell, G. Fischer, K. Konovalenko, Ground-based and space-based observations of planetary lightning. Space Sci. Revs. (2008). doi:10.1007/s11214-008-9366-8

Chapter 2
Fair-Weather Atmospheric Electrification on Earth

Abstract The electrical processes acting in Earth's atmosphere away from thunderstorms are described. The concept of a global electrical circuit is introduced, and the general conditions for a global circuit defined.

2.1 The Global Electric Circuit

If the Earth's conductive ionosphere and surface are assumed to be spherical equipotentials, a potential difference of ~ 300 kV exists between them, with the ionosphere the more positive. The ionospheric conductivity is 10^{-7}–10^{-3} Sm^{-1}, and the surface conductivity $\sim 10^{-7}$–10^{-2} Sm^{-1} (Rycroft et al. 2008). The ionosphere-surface system has a finite capacitance with an RC time constant of a few minutes, indicating that there must be a constant current source to maintain the electric fields observed in the atmosphere (Harrison and Carslaw 2003). Thunderstorms supply a net current by passing positive charge to the conductive upper atmosphere, and currents also flow from the surface up to the thundercloud (Rycroft et al. 2000). Charge exchange from particle interactions (generally between ice crystals and soft hail (graupel)) within a cumulonimbus (thunder) cloud is followed by charge separation from gravitational settling of negatively charged particles to form a dipole within the cloud. When the electric field exceeds the breakdown voltage of air, the cloud is discharged by lightning, either to the ground, within the cloud, or to other clouds (MacGorman and Rust 1998). Electrical discharges directed upwards from the thundercloud to the ionosphere, collectively called transient luminous events (TLEs) also exist (e.g. Rycroft et al. 2000, 2012). Electrostatic discharge and precipitation from electrified shower clouds comprise the disturbed weather part of the global circuit (Harrison 2005). Figure 2.1 shows a diagram of the terrestrial global circuit.

K. L. Aplin, *Electrifying Atmospheres: Charging, Ionisation and Lightning in the Solar System and Beyond*, SpringerBriefs in Astronomy, DOI: 10.1007/978-94-007-6633-4_2,

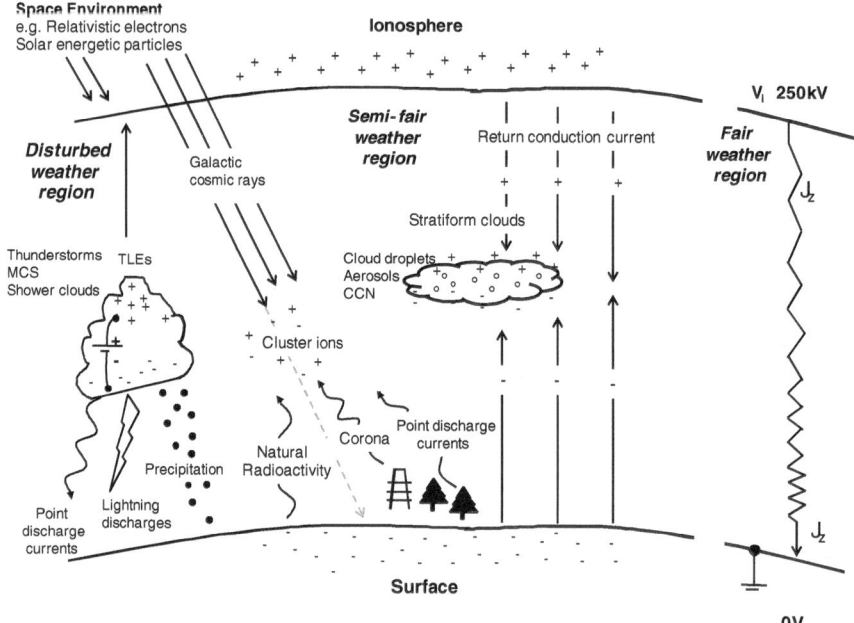

Fig. 2.1 Major processes in the terrestrial global electric circuit, after Rycroft et al. (2012). Charge separation in thunderstorms in disturbed weather regions creates a substantial potential difference between the highly conducting regions of the ionosphere and the Earth's surface. The positive potential of the ionosphere (with respect to the Earth's surface) is distributed to fair weather and semi-fair weather regions, where a small current (of density J_z) flows vertically. When this current flows through clouds it generates charge near the upper and lower cloud edges, which can influence cloud microphysical processes. (In this diagram, Mesospheric Convective Systems, which are large scale thunderstorms late in their evolution and which favour sprite generation above them, are indicated by MCS; sprites are one example of Transient Luminous Events (*TLEs*); Cloud Condensation Nuclei are shown as *CCN*)

The total current ~ 2 kA, averaged over the surface area of the Earth, gives a vertical current density, often referred to as the conduction current, $J_z \sim 2$ pAm^{-2}, carried by atmospheric ions, and an associated total global atmospheric resistance $R_T \sim 230\Omega$. Outside thunderstorm regions, the ionosphere-surface potential difference leads to a surface potential gradient PG ~ 150 Vm^{-1}. These conditions, away from charge separation processes, are often referred to as "fair weather". This paper follows the convention of defining PG (F) as dV/dz, (where z is altitude) whereas the electric field $E = -dV/dz$, so that in fair weather the PG is positive (Harrison and Carslaw 2003). PG and J_z are related to the air's electrical conductivity σ (discussed in Sect. 2.2) by

$$F = J_z/\sigma. \tag{1}$$

Typical fair weather conditions show considerable spatial variability; in particular, they are modulated by aerosol affecting the air conductivity, some of which arises from anthropogenic pollution (Harrison and Aplin 2003; Aplin 2012).

The PG variation in clean air shows a characteristic Universal Time (UT) diurnal variation arising from the integrated thunderstorm activity on each continent. This variation was first detected on the cruises of the geophysical research ship, *Carnegie*, in the first half of the twentieth century (Harrison 2012). The eponymous Carnegie variation has a broad peak, caused by the African thunderstorms dominating in the UT afternoon, and the early evening peak is related to the afternoon thunderstorm maximum in the Americas. The PG at clean air sites over land, where local aerosol concentrations are small and steady can also show the Carnegie variation (Harrison 2004). The terrestrial Carnegie variation will be compared to a Martian diurnal electric field variation in Chap. 4.

2.2 Ionisation and Atmospheric Conductivity

Terrestrial air conductivity σ arises principally from ionisation by cosmic rays, and is given by

$$\sigma = e \int n\mu \, d\mu \qquad (2)$$

where n is the mean bipolar ion concentration and μ is the mean ion mobility. The volumetric atmospheric ion production rate q is latitudinally modulated by geomagnetic deflection; the magnetic field generated by Earth's core acts like a cosmic ray energy spectrometer, permitting low energy cosmic rays to enter the atmosphere at high latitudes. Cosmic ray ionisation rates are therefore greater at higher latitudes. The solar cycle also modulates ionisation; at solar maximum the solar wind deflects more cosmic rays away from Earth, so cosmic ray ionisation varies in antiphase with the 11-year solar cycle. Ionisation rate increases from the surface (~ 10 cm^{-3}s^{-1}, ~ 80 % from natural radioisotopes emanating from the ground) up to the upper troposphere or lower stratosphere where the cosmic ray ionisation is strongest, 350 cm^{-3}s^{-1} (Makino and Ogawa 1985). Typical surface air σ values on Earth are 10^{-14}–10^{-15} Sm^{-1}.

In composition, terrestrial atmospheric ions typically consist of a central charged molecule stabilised by clustering with several ligands, commonly water or ammonia (e.g. Aplin 2008). Free electrons rapidly attach to electrophilic species to form negative ions. Atmospheric ions are lost by recombination and attachment to aerosol particles. The rate of change of bipolar ion number concentration $n_{\pm}$ can be represented as a balance between q, and two loss terms: self-recombination with coefficient α and attachment to a monodisperse aerosol population of number concentration Z and an aerosol size-dependent attachment coefficient β. The steady-state ion balance equation can be written as

$$\frac{dn_\pm}{dt} = q - \alpha n_+ n_- + \beta n_\pm Z \tag{3}$$

Once ions attach to aerosol particles, they are no longer electrically mobile enough to contribute to the atmospheric conduction current, or substantially to the air conductivity. However, there is evidence that charged aerosol can influence cloud microphysics by scavenging, the removal of aerosol by water droplets. Scavenging may be enhanced in certain regions by charging, which could affect the local cloud condensation nucleus (CCN) population or phase transitions (Tinsley et al. 2001; Tripathi and Harrison 2002). It has also been suggested that atmospheric ions can nucleate ultrafine aerosol particles, which would lead to an additional loss term in Eq. 3 (Aplin and Harrison 1999). There are two important atmospheric ion nucleation mechanisms. First, direct condensation onto ions is the process by which the ionising tracks of radioactive particles become visible in cloud chambers. In this paper, the direct nucleation mechanism is referred to as the "Wilson" mechanism, or ion-induced nucleation. This mechanism requires much larger supersaturations than can be reached in the terrestrial atmosphere (Yu and Turco 2001), but may be possible in the atmospheres of Venus (Chap. 3), Neptune (Chap. 7) and Triton (Chap. 8). Secondly, ion-mediated nucleation mechanisms have been postulated, in which charge indirectly enhances particle growth processes. Models predict ion-mediated nucleation in Earth's atmosphere (Yu and Turco 2001; Lovejoy et al. 2004), with surface, laboratory and free tropospheric observations providing evidence for the effect (Kirkby et al. 2011; Harrison and Aplin 2001; Eichkorn et al. 2002; Laakso et al. 2004), but modelling studies indicate that cloud condensation nuclei generated from ions contribute only negligibly to climate (Pierce and Adams 2009). However, there is evidence for a host of additional small effects that could link atmospheric ionisation to climate. These include the modification of droplet charges (and therefore potentially their activation) at the base of clouds (Harrison et al. 2011), and direct absorption of infrared radiation by the bending and stretching of hydrogen bonds inside atmospheric cluster-ions (Aplin and Lockwood 2013).

References

K.L. Aplin, Composition and measurement of charged atmospheric clusters. Space Sci. Rev. **137**(1–4), 213–224 (2008). doi:10.1007/s11214-008-9397-1

K.L. Aplin, Smoke emissions from industrial western Scotland in 1859 inferred from Lord Kelvin's atmospheric electricity measurements. Atmos. Env. **50**, 373–376 (2012). doi:10.1016/j.atmosenv.2011.12.053

K.L. Aplin, R.G. Harrison, The interaction between air ions and aerosol particles in the atmosphere. Inst. Phys. Conf. Series **163**, 411–414 (1999)

K.L. Aplin, M. Lockwood, Cosmic ray modulation of infra-red radiation in the atmosphere. Env. Res. Letts. **8**, 015026 (6pp) (2013). doi:10.1088/1748-9326/8/1/015026

S. Eichkorn, S. Wilhelm, H. Aufmhoff, K.H. Wohlfrom, F. Arnold, Cosmic ray-induced aerosol formation: First observational evidence from aircraft-based ion mass spectrometer measurements in the upper troposphere. Geophys. Res. Lett. **29**(14), 1698 (2002). doi:10.1029/2002GL015044

R.G. Harrison, Long term measurements of the global atmospheric electric circuit at Eskdalemuir, Scotland, 1911–1981. Atmos. Res. **70**(1), 1–19 (2004). doi:10.1016/j.atmosres.2003.09.007

R.G. Harrison, The global atmospheric electrical circuit and climate. Surv. Geophys. **25**(5–6), 441–484 (2005). doi:10.1007/s10712-004-5439-8

R.G. Harrison, The Carnegie curve. Surv. Geophys. **34**(2), 209–232 (2012). doi:10.1007/s10712-012-9210-2

R.G. Harrison, K.L. Aplin, Atmospheric condensation nuclei formation and high-energy radiation. J. Atmos. Sol.-Terr. Phys. **63**(17), 1811–1819 (2001). doi:10.1016/S1364-6826(01)00059-1

R.G. Harrison, K.L. Aplin, Nineteenth century Parisian smoke variations inferred from atmospheric electrical observations. Atmos. Env. **37**(38), 5319–5324 (2003). doi:10.1016/j.atmosenv.2003.09.042

R.G. Harrison, K.S. Carslaw, Ion-aerosol-cloud processes in the lower atmosphere. Rev. Geophys. **41**, 3 (2003). doi:10.1029/2002RG000114

R.G. Harrison, M.H.P. Ambaum, M. Lockwood, Cloud base height and cosmic rays. Proc. Roy. Soc. A. **467**(2134), 2777–2791 (2011). doi:10.1098/rspa.2011.0040

J. Kirkby, J. Curtius, J. Almeida et al., Role of sulphuric acid, ammonia and galactic cosmic rays in atmospheric aerosol nucleation. Nature **476**, 7361 (2011). doi:10.1038/nature10343

L. Laakso, T. Anttila, K.E.J. Lehtinen, P.P. Aalto, M. Kulmala, U. Hõrrak, J. Paatero, M. Hanke, F. Arnold, Kinetic nucleation and ions in boreal forest particle formation events. Atmos. Chem. Phys. **4**, 2353–2366 (2004). doi:10.5194/acp-4-2353-2004

E.R. Lovejoy, J. Curtius, K.D. Froyd, Atmospheric ion-induced nucleation of sulfuric acid and water. J. Geophys. Res. **109**, D08204 (2004). doi:10.1029/2003JD004460

D.R. MacGorman, W.D. Rust, *The Electrical Nature of Storms* (Oxford University Press, Oxford, 1998)

M. Makino, T. Ogawa, Quantitative estimation of global circuit. J. Geophys. Res. **90**, D4, 5961–5966 (1985). doi:10.1029/JD090iD04p05961

J.R. Pierce, P.J. Adams, Can cosmic rays affect cloud condensation nuclei by altering new particle formation rates? Geophys. Res. Letts. **36**, L09820 (2009). doi:10.1029/2009GL037946

M.J. Rycroft, S. Israelsson, C. Prince, The global atmospheric electric circuit, solar activity and climate change. J. Atmos. Sol-Terr. Phys. **62**, 1563–1576 (2000). doi:10.1016/S1364-6826(00)00112-7

M.J. Rycroft et al., An overview of Earth's global electric circuit and atmospheric conductivity. Space Sci. Revs. **137**, 83–105 (2008). doi:10.1007/s11214-008-9368-6

M.J. Rycroft et al., Global electric circuit coupling between the space environment and the troposphere. J. Atmos. Sol-Terr. Phys. **90–91**, 198–211 (2012). doi:10.1016/j.jastp.2012.03.015

B.A. Tinsley, R.P. Rohrbaugh, M. Hei, Electroscavenging in clouds with broad droplet size distributions and weak electrification. Atmos. Res. **59–60**, 115–135 (2001). doi:10.1016/S0169-8095(01)00112-0

S.N. Tripathi, R.G. Harrison, Enhancement of contact nucleation by scavenging of charged aerosol. Atmos. Res. **62**, 57–70 (2002). doi:10.1016/S0169-8095(02)00020-0

F. Yu, R.P. Turco, From molecular clusters to nanoparticles: Role of ambient ionisation in tropospheric aerosol formation. J. Geophys. Res. **106**(D5), 4797–4814 (2001). doi:10.1029/2000JD900539

Chapter 3
Venus

Abstract Venus's cloudy, thick, atmosphere is predicted to contain complicated distributions of charged aerosol particles, driven by cosmic ray ionisation. There are some inconsistencies between the predictions based on ion-aerosol physics, and the efficient charging required to generate lightning.

There has been much debate over the existence of lightning in Venus' dense, hot carbon dioxide atmosphere, covered with sulphuric acid clouds. This discussion was summarised in several recent review papers, in particular Yair (2012) and Yair et al. (2008), and it will not be reiterated here except where relevant. The briefest summary is essentially that the most convincing evidence for lightning on Venus is based on the magnetometer observations on Venus Express, reported by Russell et al. (2007), although there are many open questions, particularly related to the cloud electrical properties, some of which will be mentioned below.

3.1 Ionisation

Just as for Earth, solar UV radiation cannot penetrate into the lower atmosphere of Venus, and cosmic rays are the principal source of ionisation at altitudes <60 km. Borucki et al. (1982) calculated cosmic ray ionisation rates for Venus based on a model verified in the terrestrial atmosphere, Fig. 3.1 (O'Brien 1971; Borucki et al. 1982). Surface ion production rates were ~ 0.01 ion pairs $cm^{-3}s^{-1}$, rising to ~ 0.5 $cm^{-3}s^{-1}$ at 30 km (1 MPa). Ionisation at 1,000 hPa (50 km) was similar to terrestrial surface ionisation from cosmic rays. Venus has no magnetic field; thus, there will be no latitudinal effects on cosmic ray penetration into its atmosphere (Lewis 1997). However, its proximity to the Sun causes a greater modulation in ionisation rates over the solar cycle than at Earth.

Borucki et al. (1982) argued that contributions to ionisation from radioactive rocks in the ground were likely to be insignificant outside a thin layer near the

K. L. Aplin, *Electrifying Atmospheres: Charging, Ionisation and Lightning in the Solar System and Beyond*, SpringerBriefs in Astronomy, DOI: 10.1007/978-94-007-6633-4_3,

13

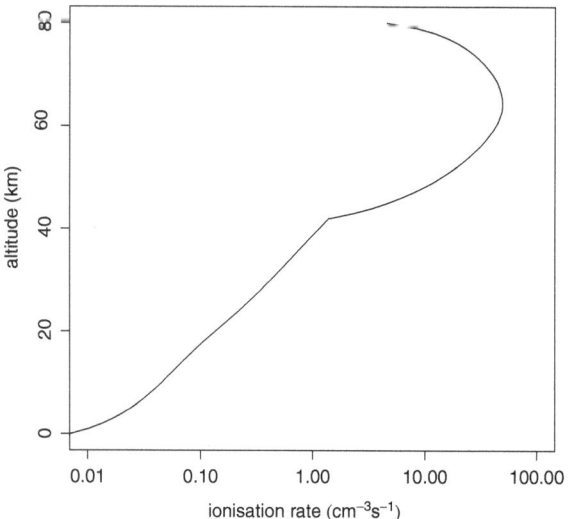

Fig. 3.1 Calculated ionisation rate profile for Venus, after Borucki et al. (1982)

surface, on the basis that the upwards penetration of radioactive particles from the ground is related to atmospheric density. Scaling arguments were used to suggest that this contribution was confined to the lowest 100 m, and cosmic rays were the only ionisation source considered in the model. However, radioactive uranium, thoron and potassium isotopes were detected at the Venusian surface by the Venera 8 mission γ-ray spectrometer in the 1970s (Vinogradov et al. 1973). The proportions of radioactive elements in the rock were similar to those found in the more radioactive terrestrial rocks like granite. The contribution of natural radioactivity to ionisation rates in the lowest 100 m of the Venusian atmosphere can be estimated by assuming that the concentration of radioactive particles emitted from the ground is the same as on Earth. Composition differences between the Venusian and terrestrial atmospheres have a negligible effect on ionisation, as the energies needed to produce an ion pair in carbon dioxide and air are similar: 33.5 and 35 eV, respectively (Wilkinson 1950; Borucki et al. 1982). The surface terrestrial ionisation rate due to natural radioactivity, ~ 8 ion pairs $cm^{-3}s^{-1}$, can be scaled by the ratio of surface air pressures on Venus and Earth to give a figure for the surface of Venus ~ 0.01 ion pairs $cm^{-3}s^{-1}$. This would double the modelled ionisation rate in the 100 m closest to the surface in comparison to what is shown in Fig. 3.1.

3.2 Ion-Aerosol Interactions

Borucki et al. (1982) modelled ion clustering reactions and interactions with a monodisperse population of aerosol particles, in what was probably the first attempt to investigate extraterrestrial ion-aerosol interactions. This work was subsequently updated by Michael et al. (2009) who used more recent input data and were able to employ modern computational techniques to extend the aerosol size distributions and charging calculations. Michael et al. (2009) carried out a detailed comparison of their work with the earlier paper by Borucki et al. (1982), and any differences between the two studies are well-understood. This chapter will summarise the methods and findings presented in both papers.

In any dense atmosphere, primary ions and electrons rapidly form ion clusters. On Venus the initial clusters are formed of carbon dioxide, which then quickly react with atmospheric trace gases in the warm conditions. Some of the final clusters formed are similar to those in the terrestrial atmosphere, particularly the hydrated cluster-ion, $H_3O^+(H_2O)_n$ (n = 3 or 4). Other common species are $H_3O^+(SO_2)$ and $H_3O(H_2O^+)(CO_2)$; the average positive ion mass is $\sim 80 \pm 40$ amu. Less chemical information is available with which to model negative ion evolution, but sulphate species appear to dominate, with a mean negative ion mass $\sim 150 \pm 75$ amu. Free electrons are also present above 60 km, so there are three ion balance equations for Venus, with electrons and negative ions as two distinct species. The term representing positive ions is similar to the terrestrial case (Eq. 2), except that there are two recombination terms, one for negative ions and one for electrons; a similar approach can be used to model ion–electron physics in the terrestrial ionosphere (Ratcliffe 1972). There is an additional loss term in the equation for negative ions representing electron detachment by collision, attachment and photo-induced mechanisms. The third ion balance equation, representing the rate of change of electron concentration, includes attachment terms for positive ions and aerosols, and an electron detachment term. A term to represent the attachment to neutral oxygen and sulphur dioxide molecules was included explicitly for negative ions and electrons.

Because of the ubiquity of cloud cover on Venus, attachment to aerosol particles is an important global ion loss mechanism. Venusian clouds exist from ~ 50–70 km and are composed principally of sulphuric acid (James et al. 1997). In the Borucki et al. (1982) model, a trimodal aerosol size distribution based on data from the Pioneer and Venera probes was assumed. In Michael et al. (2009), three cloud layers were assumed, with the lower cloud layer having a trimodal particle distribution and the others considered as bimodal. The particle size distributions were taken from James et al. (1997). This is a substantial improvement upon the simple assumptions made by Borucki et al. (1982) on the basis of limited data. In both modelling studies, attachment coefficients were calculated using a simplified parameterisation for each mode, in each atmospheric layer. Ion and electron mobilities were obtained using the McDaniel and Mason (1973) method, whereas ion and electron concentrations were calculated by solving the three ion balance equations in the steady state. Solving the set of differential equations across each particle size allowed Michael et al. (2009) to predict the charge distribution on cloud particles in the three cloud layers, summarised in Fig. 3.2. The existence of mobile free electrons in the upper cloud layers means that high negative charges can be acquired by the cloud particles, whereas in the lower cloud layers the

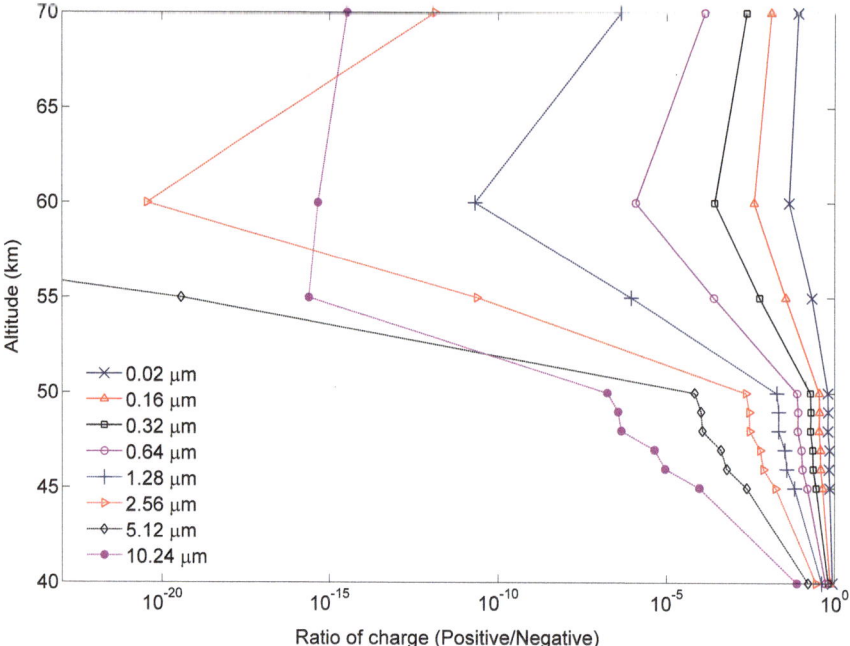

Fig. 3.2 Ratio of total positive to negative charges in the Venusian cloud layers, over a range of particle sizes. Reproduced with permission from Michael et al. (2009)

particles are on average more positively charged due to the greater mobility of positive Venusian ions (Aplin 2008).

Venusian air conductivity is generally similar to that on Earth at similar pressures, with a global reduction of a factor of 2–6 due to the cloud layer (Michael et al. 2009). Using a polydisperse particle distribution reduces attachment overall compared to a monodisperse distribution, and therefore the effect of cloud particles on the conductivity is smaller for the more realistic polydisperse case.

The cloud charging predicted by Michael et al. (2009) has interesting consequences for lightning generation. Figure 3.2 indicates that the top cloud layer, particularly the larger particles, is many orders of magnitude more negatively charged than the middle and bottom cloud layers. This charge separation could start the build-up of an electric field that could lead to lightning. As in terrestrial thunderstorms, mechanical processes such as convection can also assist in separating large particles from small ones, if leakage currents are low enough to permit the particles to sustain their charge. In an order of magnitude calculation, Michael et al. (2009) suggest that the air is conductive enough for the cloud charge to leak away more quickly than electric fields can be generated, which would prevent lightning even in the case of extremely efficient charging.

3.3 Ion-Induced Nucleation

In a detailed microphysics model, James et al. (1997) identified two cloud formation mechanisms. At the top of the cloud layer, particles are photochemically produced, but near the bottom, cloud formation is thought to result from heterogeneous nucleation. The atmosphere is supersaturated with respect to sulphuric acid from 40 km upwards, where H_2SO_4 vapour can condense onto hydrated sulphuric acid particles, similar to terrestrial stratus generation by condensation of water vapour onto cloud condensation nuclei. As described in Sect. 3.2, Venusian atmospheric ions are likely to be sulphuric acid hydrates. The appropriate conditions may therefore exist for aerosols to form by ion-induced nucleation of sulphuric acid. The supersaturation required for ions to grow into ultrafine droplets by direct condensation can be determined using the Thompson equation (Mason 1971). This equation describes the equilibrium saturation ratio needed for ion-induced nucleation to become energetically favourable. The equilibrium condition is defined at a saturation ratio S, where r is radius, ρ fluid density, M the mass of the molecule, q charge, γ_T the surface tension, k_B Boltzmann's constant, T temperature, r_o the initial radius (all in SI units), and ε_r relative permittivity:

$$\ln S = \frac{M}{k_B T \rho} \left[\frac{2\gamma_T}{r} - \frac{q^2}{32\pi^2 \varepsilon_0 r^4} \left(1 - \frac{1}{\varepsilon_r} \right) \right]. \tag{3.1}$$

Equation 3.1 can be used to assess if condensation of gaseous H_2SO_4 onto ions is likely to occur in the lower cloud-forming regions at ~ 40 km in the Venusian atmosphere (Aplin 2006). Kolodner and Steffes (1998) found that sulphuric acid supersaturations of 25–88 % were readily observed in the cloud-forming region of Venus' atmosphere. Temperature, surface tension, density and dielectric constant data were also obtained and substituted into Eq. 3.1 to compute the saturation ratio needed for nucleation onto differerently charged condensation nuclei at the different Venusian locations for which H_2SO_4 profiles were retrieved.

As discussed in Sect. 2.2, "Wilson" nucleation is impossible in the terrestrial atmosphere because water supersaturations rarely exceed a few percent, and $S = 4$ is required for condensation onto ions. Sulphuric acid has a lower vapour pressure and a higher permittivity (polarisability) than water, so the saturations required for condensation onto singly charged Venusian ions are lower than on Earth. Despite this, S is never high enough for condensation onto particles with one electronic charge. The altitudes most favourable for ion-induced nucleation are a tradeoff between temperature, supersaturation and number of charges needed (Eq. 3.1). Equation 3.1 and the measurements of sulphuric acid supersaturation presented by Kolodner and Steffes (1998) can be used to investigate the possibility of ion-induced nucleation. The most probable location and altitude for ion-induced nucleation is at 88°S and ~ 43 km, a region with relatively low SS, 25–30 % (it has been measured up to 85 % at a higher altitude) but the higher temperatures (~ 395 K), reduce the SS needed for condensation onto charged particles. This is illustrated in Fig. 3.3 which shows the solution of Eq. 3.1 at 395 K.

Fig. 3.3 Supersaturations required for Wilson nucleation of sulphuric acid onto small particles with 1, 2 and 5 electronic charges for the conditions measured at 43 km by Kolodner and Steffes (1998)

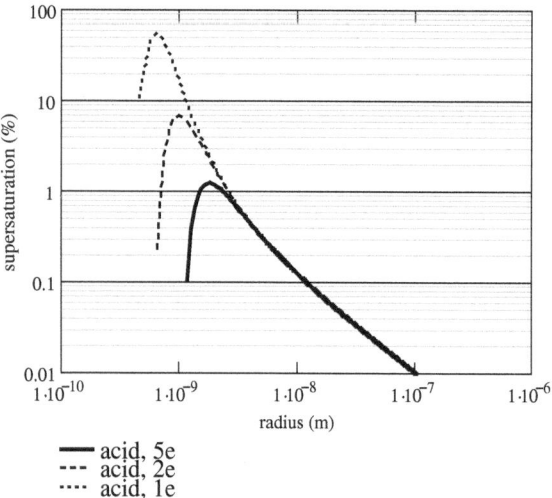

Doubly charged particles of radius 1 nm can nucleate at SS = 7 %, and particles with 5 charges of radius 2 nm can nucleate at SS = 1–2 %. These SS appear to be relatively easily attained at 88°S.

The possibility of doubly charged particles existing in this region can be estimated from the aerosol charge distribution arising from ion-aerosol attachment processes as computed by Michael et al. (2009). The nearest data they present to 43 km, where the best conditions for ion-induced nucleation are expected, are presented in Fig. 3.4. The probability of double charging of the smallest particles $\sim (10^{-8}m)$ is ~ 0.001, which is lower than the estimate obtained by Aplin (2006) based on a simple monodisperse approximation.

These estimates do not rule out the possibility of direct condensation of gaseous sulphuric acid onto charged particles contributing to Venusian cloud formation.

3.4 Is There a Global Electric Circuit on Venus?

At the moment, the existence of a Venusian global electric circuit seems unlikely due to the lack of conclusive evidence for current flow through precipitation or electrostatic discharges. Based on the assumption that there is lightning, then there is a good basis for the existence of a global circuit if the surface is more electrically conductive than its atmosphere. The composition of the Venusian surface is not yet well known because the cloudy, dense atmosphere makes remote sensing of the surface difficult. Most of the data is based on results from the three landers, Venera 13, Venera 14 and Vega 2, which all landed on a similar, basaltic, region of the planet (Lewis 1997). Radar data transmitted by the Magellan spacecraft in orbit around Venus suggested that the electrical conductivity of the Maxwell Montes highland region was $\sim 10^{-13}\,Sm^{-1}$ (Pettengill et al. 1996), and Simões et al. (2008)

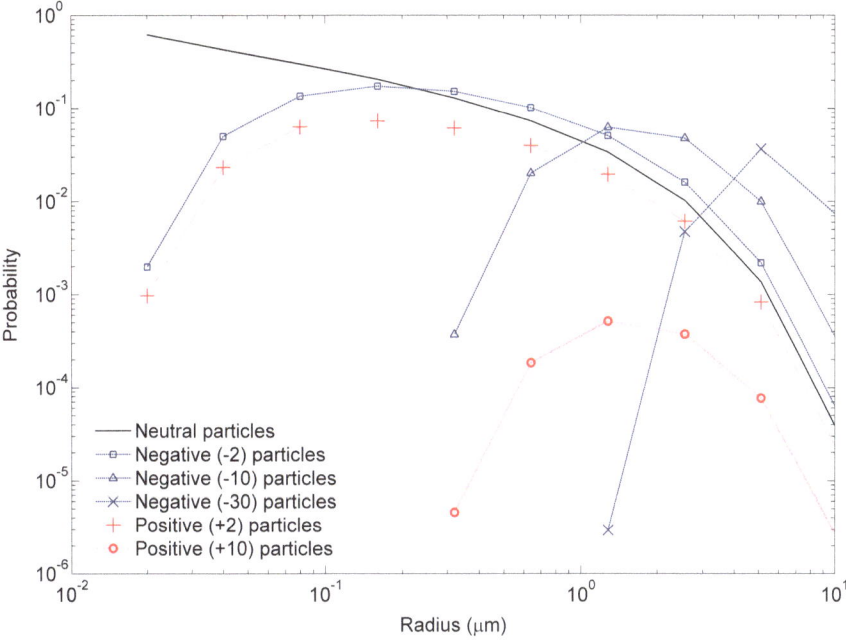

Fig. 3.4 Probability of cloud particle charging at 46 km for a range of particle sizes. Reproduced with permission from Michael et al. (2009)

estimated a surface conductivity of 10^{-6}–10^{-4} Sm^{-1}. As the average air conductivity on Venus is $\sim 10^{-14}$ Sm^{-1} then the ground does appear to be more conductive than the air. A Venusian global circuit therefore seems plausible.

3.5 Current and Future Missions

Venus Express, a European Space Agency orbiter, is currently studying the atmosphere, mapping surface temperatures and measuring the interaction of the atmosphere with the solar wind, with instrumentation principally based on the successful Mars Express mission. Whilst new atmospheric profile and composition data will undoubtedly be of use for improving models of Venusian ion-aerosol physics, the spacecraft carries little specific atmospheric electrical instrumentation, although Russell et al. (2007) have successfully used its magnetometer for lightning detection.

The proposed European Venus Explorer mission (Chassefière et al. 2009a) would answer many of the open questions about Venus atmospheric electricity not addressed by Venus Express. The mission would deploy an orbiter, descent probe and balloon (Chassefière et al. 2009b) to better understand the planet's evolution and climate through atmospheric and surface measurements. The model payload specifies electrical instrumentation for both the balloon and descent probe, with the descent probe focusing on transient events (lightning and gamma ray flashes) and the balloon's electrical antennae capable of measuring both the "fair" and "disturbed" weather properties, with a microphone to detect thunder. Other instrumentation will characterise clouds and aerosol particles, with the possibility of using a quartz crystal microbalance to measure charged aerosol directly.

References

K.L. Aplin, Atmospheric electrification in the Solar System. Surv. Geophys. **27**(1), 63–108 (2006). doi:10.1007/s10712-005-0642-9

K.L. Aplin, Composition and measurement of charged atmospheric clusters. Space Sci. Rev. **137**(1–4), 213–224 (2008). doi:10.1007/s11214-008-9397-1

W.J. Borucki, Z. Levin, R.C. Whitten, R.G. Keesee, L.A. Capone, O.B. Toon, J. Dubach, Predicted electrical conductivity between 0 and 80 km in the Venusian atmosphere. Icarus **51**, 302–321 (1982). doi:10.1016/0019-1035(82)90086-0

E. Chassefière, O. Korablev, T. Imamura et al, European Venus Explorer (EVE): an in situ mission to Venus. Exp. Astron. **23**(3), 741–760 (2009a). doi:10.1007/s10686-008-9093-x E

E. Chassefière, O. Korablev, T. Imamura et al, European Venus Explorer: an in situ mission to Venus using a balloon platform. Adv. Space Res. **44**, 106–115 (2009b). doi:10.1016/j.asr.2008.11.025

E.P. James, O.B. Toon, G. Schubert, A numerical microphysical model of the condensational Venus cloud. Icarus **129**, 147–171, art. no. IS975763 (1997). doi:10.1006/icar.1997.5763

M.A. Kolodner, P.G. Steffes, The microwave absorption and abundance of sulphuric acid vapour in the Venus atmosphere based on new laboratory measurements. Icarus **132**, 151–169 (1998). doi:10.1006/icar.1997.5887

J.S. Lewis, *Physics and chemistry of the solar system* (Academic Press, San Diego, 1997)

B.J. Mason, *Physics of Clouds*. (Pergamon, Oxford, 1971)

E.W. McDaniel, E.A. Mason, *The mobility and diffusion of ions in gases* (Wiley, New York, 1973)

M. Michael, S.N. Tripathi, W.J. Borucki, R.C. Whitten, Highly charged cloud particles in the atmosphere of Venus. J. Geophys. Res. **114**, E04008 (2009). doi:10.1029/2008JE003258

K. O'Brien, Cosmic-ray propagation in the atmosphere. Il nuovo cimento **3**(3), 521–547 (1971)

G.H. Pettengill, P.G. Ford, R.A. Simpson, Electrical properties of the Venus surface from bistatic radar observations. Science **272**(5268), 1628–1631 (1996). doi:10.1126/science.272.5268.1628

J.A. Ratcliffe, *An Introduction to the Ionosphere and Magnetosphere*. (Cambridge University Press, Cambridge, 1972)

C.T. Russell, T.L. Zhang, M. Delva et al., Lightning on Venus inferred from whistler-mode waves in the ionosphere. Nature **450**, 661–662 (2007). doi:10.1038/nature05930

F. Simões, M. Hamelin, R. Grard, K.L. Aplin, C. Beghin, J.J. Berthelier, B.P. Besser, J.P. Lebreton, J.J. Lopez-Moreno, G.J. Molina-Cuberos, K. Schingenschuh, T. Tokano, Electromagnetic wave propagation in the surface-ionosphere cavity of Venus. J. Geophys. Res. **E7**(113), E07007 (2008). doi:10.1029/2007JE003045

A.P. Vinogradov, Y.A. Surkov, F.F. Kirnozov, The content of uranium, thorium and potassium in the rocks of Venus as measured by Venera 8. Icarus, **20**(3), 253–259 (1973). doi:10.1016/0019-1035(73)90001-8

D.H. Wilkinson, *Ionisation Chambers and Counters* (Cambridge University Press, Cambridge, 1950)

Y. Yair, New results on planetary lightning. Adv. Space Res. **50**, 293–310 (2012). doi:10.1016/j.asr.2012.04.013

Y. Yair, G. Fischer, F. Simoes, N. Renno, P. Zarka, Updated review of planetary atmospheric electricity. Space Sci. Rev. **137**, 29–49 (2008). doi:10.1007/s11214-008-9349-9

Chapter 4
Mars

Abstract The Martian environment is electrostatically active, but lightning has not yet been detected. Mars is expected to have a global electric circuit, similar to its terrestrial analogue but with dust storms as the charge generator. Tests of these predictions are hindered by a lack of in situ electrical observations.

The proximity of Mars to Earth has meant that meteorological and geophysical data have been acquired from several space missions. Unfortunately none has included electrical instrumentation, and the only experimental evidence for electrical activity on Mars is the electrostatic adhesion of dust to the wheels of the Mars Pathfinder and Sojourner rovers (Farrell et al. 1999; Ferguson et al. 1999; Berthelier et al. 2000). However, based on deductions from the terrestrial analogues in desert sandstorms, Mars is expected to have substantial atmospheric electrification.

Mars is unique in the solar system because of the significance of dust in its climate. The fine dust grains on the surface can be suspended by wind to form dust storms which can sometimes cover large areas of the planet. Convective vortices, often analogous in structure and size to terrestrial dust devils, but also commonly an order of magnitude larger (Balme and Greeley 2006), play an important part in lofting dust into the atmosphere. One consequence of the dust's low thermal inertia (Lewis 1997) is that the surface responds rapidly to solar radiation, leading to large diurnal variations in temperature. Atmospheric dust loading could have important radiative effects on Mars; meteorological models developed from terrestrial general circulation models are being used to investigate the role of dust in Martian global climate (Newman et al. 2002), and temperatures in the Martian northern hemisphere spring and summer could be controlled by convectively lofted dust (Basu and Richardson 2004).

K. L. Aplin, *Electrifying Atmospheres: Charging, Ionisation and Lightning in the Solar System and Beyond*, SpringerBriefs in Astronomy,
DOI: 10.1007/978-94-007-6633-4_4,

4.1 Ionisation and Atmospheric Conductivity

Unlike Earth, Mars does not have an internal dynamo generating a global magnetic field with which to deflect cosmic rays away from the planet. Instead, the Martian magnetic field arises from magnetisation of the Martian crust. The Mars Global Surveyor instrument carried a magnetometer, from which global maps of the Martian magnetic field were produced (Connerney et al. 2001). It indicated that most magnetisation was concentrated in the southern hemisphere with maximum radial fields reaching 200nT (mapped from a height of 400 km) in the Terra Cimmeria/Terra Sirenum region, centred on $\sim 60°S$, 180°. The Martian cosmic ray flux is likely to be higher than Earth's in its less dense atmosphere, with no latitudinal modulation, but the regional magnetic field might modulate ionisation rates to be slightly higher in the northern hemisphere. In the daytime, solar UV photons can penetrate to the surface, as, unlike Earth, there is no atmospheric ozone layer to absorb them (Fillingim M., Global electric circuit of Mars, Florida Institute of Technology, available at http://sprg.ssl.berkeley.edu/$\sim$matt/mars/mars_gec.html, Unpublished report, 1998). They ionise both the surface and atmosphere to leave the surface slightly positively charged, with the emitted photoelectrons dominating the daytime atmospheric conductivity (Grard 1995). Typical surface daytime free electron densities are 1–100 cm^{-3}, dropping to <1 cm^{-3} at night. The free electrons and ions formed by electron impact ionisation cause high air conductivity at the surface during the day, $\sim 10^{-11}$ Sm^{-1}, comparable to terrestrial stratospheric conductivity (Berthelier et al. 2000). Michael et al. (2008) calculated the effect of aerosols (dust) on the conductivity with an electron–ion–aerosol model using up-to-date data and found that surface conductivity varied from 4×10^{-12} to 3×10^{-11} Sm^{-1} from an nocturnal aerosol-laden atmosphere (e.g. during a dust storm) to a clean, daytime atmosphere.

The daytime positive conductivity is small because although there are fewer free electrons than positive ions by two orders of magnitude, electron mobility is about three orders of magnitude greater than positive ion mobility, so the electron conductivity dominates (Michael et al. 2008). It is not clear whether the atmospheric electronegative species such as oxygen will react with all the free electrons produced by GCR ionisation at night (Molina-Cuberos et al. 2002); if there are no free electrons (as stated by Michael et al. 2008) then the conductivity will be more evenly balanced between positive and negative ions. The large diurnal variation in atmospheric conductivity from photoionisation is estimated to be a factor of 2.5 (Michael et al. 2008), and will be discussed in Sect. 4.4.

4.2 Electrical Discharges

Electrical discharges have not yet been detected on Mars, but it is widely supposed that they occur. Up to 10^6 elementary charges cm^{-3} (0.1 $pCcm^{-3}$) and electric fields of kVm^{-1}, generated by triboelectric charging, have been measured in terrestrial dust devils (Farrell and Desch 2001). Calculations and laboratory measurements indicate that high electric fields are also expected in Martian dust devils, although not all of these laboratory results are applicable to the Martian environment (Aplin et al. 2012). Farrell et al. (1999) suggest that electric fields generated in a dust devil are likely to be limited by the breakdown voltage of the low-pressure Martian atmosphere, $\sim 10 \text{kVm}^{-1}$, rather than the maximum charge sustained by an individual dust grain. They predicted that a dust devil 5 km across could sustain ≤ 200 elementary charges cm^{-3} (32 aCcm^{-3}) before breakdown. Two breakdown mechanisms were suggested, firstly a corona or glow discharge due to local breakdown of the air at the edge of the dust devil, which self-limits the maximum field attained, and secondly a spark discharge similar to terrestrial volcanic lightning (Farrell et al. 1999; Farrell and Desch 2001).

Attempts have been made to detect lightning remotely using spacecraft radio measurements. Ruf et al. (2009) attributed changes in the non-thermal microwave emissions from a Martian dust storm to Schumann resonances excited by lightning, but so far this work has not been corroborated. In a thorough study Gurnett et al. (2010) searched for transient radio signals between 4.0 and 5.5 MHz in five years' worth of data, with no positive results. In situ measurements may be required, to be discussed in Sect. 4.5.

4.3 Dust-Driven Global Circuit

Farrell and Desch (2001) have discussed the possibility of a Martian atmospheric electrical circuit with electrical discharges in dust storms as current generators, similar to thunderstorms on Earth. As defined in Chap 2, a global circuit requires the existence of a conductive ionosphere and surface, charge separation and a conductive atmosphere. The Martian ionosphere extends upwards from ~ 15 km, much lower than on Earth because of the presence of free electrons in the lower atmosphere (Berthelier et al. 2000). In their model Farrell and Desch (2001) assumed that the ground conductivity $>10^{-9} \text{ Sm}^{-1}$. Based on the assumption that Martian charge separation processes in dust clouds are not as efficient as theoretical considerations suggest, as for terrestrial dust devils, they estimated the "fair weather" electric field and conduction current density generated by dust storms. In the Martian context, "fair weather" refers to areas not covered by a dust storm. In the storm season (northern hemisphere winter) one very large regional ($500 \times 500 \times 20$ km) storm is expected, with several medium-sized ($50 \times 50 \times 15$ km) storms. The resistance in the air column above the dust storm was estimated from electron density models, and the voltage in the dust cloud was assumed to be 1kVm^{-1} multiplied by the cloud height. Ohm's Law was then used to calculate the current contributed to the global circuit: $\sim 2 \text{kA}$ for the regional storm and $\sim 500\text{A}$ in total from the moderate storms. This current is averaged over the fair weather area to give an electric field $E = 475 \text{ Vm}^{-1}$ and a conduction current of $J_z = 1.3 \text{ nAm}^{-2}$. Outside the storm season, dust devils at the surface are common, and the only source of charge for the global circuit.

Fig. 4.1 Schematic of a possible Martian global electric circuit, based on Farrell and Desch (2001)

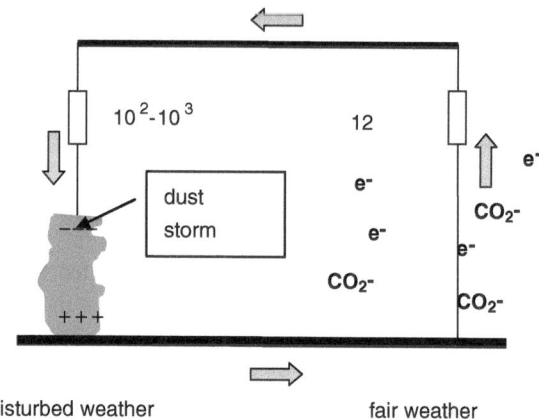

If each is estimated to contribute 1A then this gives $E = 0.14$ Vm^{-1} and $J_z = 0.4$ pAm^{-2}. The daytime fair weather electric field may be enhanced by 0.01–0.1 Vm^{-1} by the emission of photoelectrons from the surface (Grard 1995). The aerosol-free atmospheric conductivity calculated using these estimates from Eq. 1.1, is consistent with the estimates provided by Michael et al. (2008). A schematic diagram of the model of Farrell and Desch (2001) is shown in Fig. 4.1. Note that the current flow and electric fields are oppositely directed to those in the terrestrial global circuit, Fig. 4.1.

In the absence of direct measurements, there are substantial uncertainties in the Farrell and Desch (2001) model. Discharge via local corona currents would not contribute to the global circuit, unlike sparks to the ground or upper atmosphere. The likelihood of local corona discharge is dependent on the efficiency of the tribocharging process; if it is more efficient than assumed by Farrell and Desch (2001), coupling to the global circuit will drop, reducing E and J_z. The conductivity of the Martian surface is also poorly understood; values in the literature from 10^{-7} to 10^{-11} Sm^{-1} were discussed by Berthelier et al. (2000) who believed the best estimate was 10^{-10} to 10^{-12} Sm^{-1}. The ratio of ground to air conductivity σ_g/σ_a is 10^4 to 0.1: according to the constraints defined in Chap. 1, σ_g/σ_a must be ≥ 1 for a global circuit (on Earth $\sigma_g/\sigma_a \sim 10^9$). If the atmospheric conductivity is $\sim 10^{-11}$ to 10^{-12} Sm^{-1} then a Martian global circuit seems less likely; however, both atmospheric and surface conductivity are poorly constrained due to lack of data.

4.4 Variability in Martian Atmospheric Electricity

Martian fair weather atmospheric electricity is likely to be much more variable than on Earth, for reasons first summarised by Fillingim, (Unpublished report, 1998). Diurnal variations are substantial, and photoelectrons dominate the Martian daytime atmospheric conductivity because of their high number concentration and electrical mobility, estimated to be ~ 10 m^{-2}V^{-1}s^{-1} for typical Martian surface conditions, compared to ~ 0.02 m^{-2}V^{-1}s^{-1} for ions (Borucki et al. 1982; Pack et al. 1962). (Note that the ion mobility is much greater than that at Earth's surface, because of the low-pressure atmosphere). At night there is no photoelectron emission, and charged particles are only formed by cosmic ray ionisation, causing a large diurnal air conductivity variation. The air density on Mars varies by 20 % over the year (Lewis 1997), leading to seasonal variations in cosmic ray penetration into the atmosphere. The daytime air conductivity near the surface is unlikely to be significantly affected by the seasonal pressure variation because of the dominance of photoelectrons.

Diurnal and spatial variations in the storms contributing charge to the global circuit modulate the daily cycle of fair weather electric field, described in Chap 5. If a global circuit exists on Mars then the planet may have its own "Carnegie" diurnal electric field variation. This depends on both the location of the dust storms contributing to the global circuit and their diurnal variation. If these regions and variations can be identified, then it is possible to predict a Martian "Carnegie curve" using data on the diurnal variation of dust devils, and the distribution of large dust storms. Ringrose et al. (2003) have analysed the Viking 2 lander meteorological data from 1976, and identified convective vortices passing near the lander. From this they present a diurnal variation in dust devil activity peaking at about 1300 local time, Fig. 4.2a. This is similar to the pattern of terrestrial

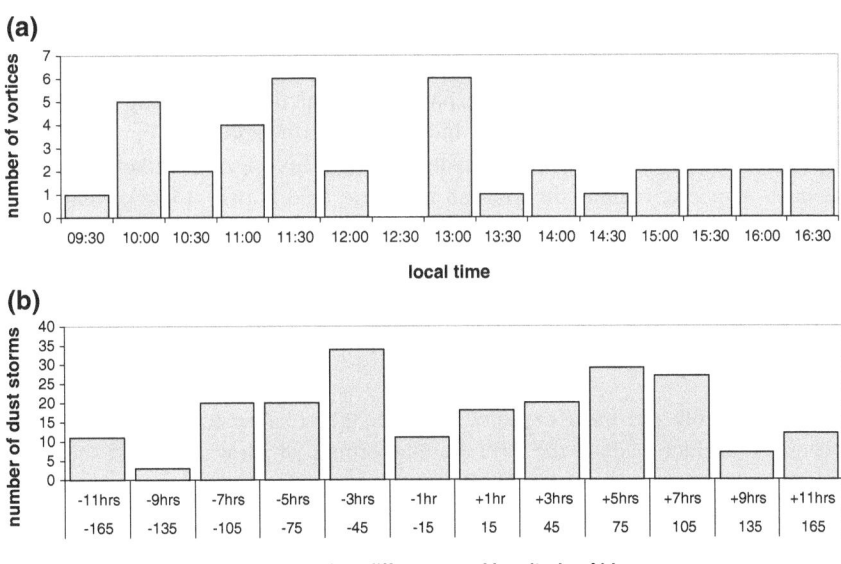

Fig. 4.2 a Diurnal statistics of convective vortices seen by the Viking 2 lander, after Ringrose et al. (2003). **b** Onset locations of large Martian dust storms, binned in 30° bands of longitude, after Newman et al. (2002). The time difference from the time at 0° is also indicated

convective vortex generation, with an unexpected morning peak. The starting locations of large Martian dust storms from 1894 to 1990 have been plotted in Newman et al. (2002); these data are presented in simplified form in Fig. 4.2b, showing the number of storms in 30° longitudinal bins. The diurnal variation in relative intensity of atmospheric electric field at 0° longitude has been computed based on the dust storm diurnal variation at different Martian local times, to derive a Martian "Carnegie curve" (Aplin 2006). It shows a bimodal pattern with a nocturnal minimum. This is a consequence of the lack of dust devils at night, and the scarcity of storms at longitudes with a 10–12 h time difference from 0°. The morning peak results from the dust storm activity around Hellas Planitia at 60–130° E, and the afternoon peak from the strongest single dust storm area, in the Xanthe Terra/Marineris region (Greeley and Batson 2001). The shape of the curve was relatively insensitive to the magnitude of the morning peak in dust devil activity, which may be of instrumental origin (Ringrose et al. 2003). The Martian "Carnegie curve" is an estimate based on the limited data available, and assumes that all Martian dust storms, irrespective of size, have the same diurnal variation as dust devils, and the spatial variation of the starting location of large storms. All storms are also assumed to contribute equally to the global circuit. The Martian Carnegie curve is expected to have a substantial seasonal variation, as indicated by Farrell and Desch (2001), because of seasonal changes in dust storm frequency.

Mars has variable topography: its highest peak, the biggest mountain in the Solar System, Olympus Mons, is 25 km high (Lewis 1997). At Olympus Mons and the nearby Tharsis range, the distance from surface to ionosphere is substantially reduced, which should increase the local conduction current (Fillingim, Unpublished report, 1998). The variations of magnetic field over Mars are likely to have complex consequences near the more magnetic regions. Fillingim, (Unpublished report, 1998) pointed out that atmospheric thermal tides may drag ionospheric plasma across magnetic field lines, inducing electric fields and currents. Local cosmic ray penetration could be modulated, though this appears unlikely as even in the most magnetic regions the Martian magnetic field is two orders of magnitude lower than Earth's magnetic field (Molina-Cuberos et al. 2001).

4.5 Future Mission Plans

Mars is arguably the most explored body in the solar system, having received several visits since Mars 2, the 1971 Russian orbiter which was the first successful mission to the red planet (Lewis 1997). The absence of relevant instrumentation on any of the missions severely limits our understanding of the Martian electrical environment. The lack of quantitative measurements of Martian atmospheric electrification could be a constraint on future missions, especially as Mars will almost certainly be the first extraterrestrial planet to be visited by humans, and all potential hazards must be quantified before this, to minimise risk. Electrostatic measurements could also be used to improve the performance of future unmanned missions to Mars, by quantifying the hazard to robotic explorers. Pechony and Price (2004) predicted that any extremely low frequency (ELF) resonances resulting from Martian electrostatic discharges would not be well-defined, so the most promising approach for detecting atmospheric electrical activity appears to be through in situ instrumentation. Calle et al. (2004) describe electrostatic sensors embedded in the wheels of rovers, measuring the charge induced on a metal electrode by soil or dust particles. The variation in charging with the soil material could provide mineralogical as well as electrical information. Other Martian atmospheric electrical instrumentation has been proposed, such as a sensor to detect both atmospheric electric fields and conductivity (Berthelier et al. 2000), currently on the payload for the ExoMars mission scheduled for launch in 2016 (Esposito et al. 2012). A prototype spectrometer to detect the charge and size distribution of atmospheric dust has also been developed (Fuerstenau and Wilson 2004). The NASA Curiosity rover, which successfully landed on Mars in August 2012 carries an energetic particle analyser (Hassler et al. 2012) to quantify the ionisation rate and radiation environment, so this will at least start to constrain the modelling work reported above.

References

K.L. Aplin, Atmospheric electrification in the Solar System. Surv. Geophys. **27**(1), 63–108 (2006). doi:10.1007/s10712-005-0642-9

K.L. Aplin, T. Goodman, K.L. Herpoldt, C.J. Davis, Laboratory analogues of Martian electrostatic discharges. Planet. Space Sci. **69**, 100–104 (2012). doi:10.1016/j.pss.2012.04.002

M. Balme, R. Greeley, Dust devils on Earth and Mars. Rev. Geophys. **43**, RG3003 (2006). doi:10.1029/2005RG000188

S. Basu, M.I. Richardson, Simulation of the Martian dust cycle with the GFDL Mars GCME. J. Geophys. Res. **109**(E11), 11006 (2004). doi:10.1029/2004JE002243

J.J. Berthelier, R. Grard, H. Laakso, M. Parrot, ARES, atmospheric relaxation and electric field sensor, the electric field experiment on NETLANDER. Planet. Space Sci. **48**, 1193–1200 (2000). doi:10.1016/S0032-0633(00)00103-3

W.J. Borucki, Z. Levin, R.C. Whitten, R.G. Keesee, L.A. Capone, O.B. Toon, J. Dubach, Predicted electrical conductivity between 0 and 80 km in the Venusian atmosphere. Icarus **51**, 302–321 (1982)

C.I. Calle, J.G. Mantovani, C.R. Buhler, E.E. Groop, M.G. Buehler, M.G. Nowicki, Embedded electrostatic sensors for Mars exploration missions. J. Electrostat. **61**, 245–257 (2004). doi:10.1016/j.elstat.2004.03.001

J.E.P. Connerney, M.H. Acuna, P.J. Wasilewski, G. Kletetschka, N.F. Ness, H. Reme, R.P. Lin, D.L. Mitchell, The global magnetic field of Mars and implications for crustal evolution. Geophys. Res. Lett. **28**(21), 4015–4018 (2001). doi:10.1029/2001GL013619

F. Esposito, F. Montmessin, S. Debei, et al., The DREAMS payload on-board the Entry and descent Demonstrator Module of the ExoMars mission. Eur. Geophys. Union Gen. Assembly, Vienna, **22–27**, 9722 (2012)

W.M. Farrell, M.D. Desch, Is there a Martian atmospheric electric circuit. J. Geophys. Res. **E4**, 7591–7595 (2001). doi:10.1029/2000JE001271

W.M. Farrell, M.L. Kaiser, M.D. Desch, J.D. Houser, S.A. Cummc, D.M. Wilt, G.A. Landis, Detecting electrical activity from Martian dust storms. J. Geophys Res. **104**(2), 3795–3801 (1999). doi:10.1029/98JE02821

D.C. Ferguson, J.C. Kolecki, M.W. Siebert, D.M. Wilt, J.R. Matijevic, Evidence for Martian electrostatic charging and abrasive wheel wear from the Wheel Abrasion Experiment on the Pathfinder Sojourner rover. J. Geophys. Res. **104**(E4), 8747–8759 (1999). doi:10.1029/98JE02249

S. Fuerstenau, G. Wilson, A particle charge spectrometer for determining the charge and size of individual dust grains on Mars. Inst. Phys. Conf. Ser. **178**(4), 143–148 (2004). doi:10.1201/9781420034387.ch23

R. Grard, Solar photon interaction with the Martian surface and related electrical and chemical phenomena. Icarus **114**, 130–138 (1995). doi:10.1006/icar.1995.1048

R. Greeley, R. Batson, *The Compact NASA Atlas of the Solar System* (Cambridge University Press, Cambridge, 2001)

D. Gurnett, D.D. Morgan, L.J. Granroth, B.A. Cantor, W.M. Farrell, J.R. Espley, Non-detection of impulsive radio signals from lightning in Martian dust storms using the radar receiver on the Mars Express spacecraft. Geophys. Res. Lett. **37**, L17802 (2010). doi:10.1029/2010GL044368

D.M. Hassler, C. Zeitlin, R.F. Wimmer-Schweingruber et al., The radiation assessment detector (RAD) investigation. Space Sci. Revs. **170**, 503–558 (2012). doi:10.1007/s11214-012-9913-1

J.S. Lewis, *Physics and chemistry of the solar system* (Academic Press, San Diego, 1997)

M. Michael, S.N. Tripathi, S.K. Mishra, Dust charging and electrical conductivity in the day and nighttime atmosphere of Mars. J. Geophys. Res. **113**, E07010 (2008). doi:10.1029/2007JE003047

G.J. Molina-Cuberos, H. Lichtenegger, K. Schwingenschuh, J.J. López-Moreno, R. Rodrigo, Ion-neutral chemistry model of the lower ionosphere of Mars. J. Geophys. Res. **107**(E5), 5027 (2002). doi:10.1029/2000JE001447

G.J. Molina-Cuberos, W. Stumptner, H. Lammer, N.I. Kömle, K. O'Brien, Cosmic ray and UV radiation models on the ancient Martian surface. Icarus **154**, 216–222 (2001). doi:10.1006/icar.2001.6658

C.E. Newman, S.R. Lewis, P.L. Read, F. Forget, Modeling the Martian dust cycle, 1. Representations of dust transport processes. J. Geophys. Res. **107**(E12), 5123 (2002). doi:10.1029/2002JE001910

J.L. Pack, R.W. Voshall, A.V. Phelps, Drift velocities of slow electrons in krypton, xenon, deuterium, carbon monoxide, carbon dioxide, water vapor, nitrous oxide and ammonia. Phys. Rev. **127**(6), 2084–2089 (1962). doi:10.1103/PhysRev.127.2084

O. Pechony, C. Price, Schumann resonance parameters calculated with a partially uniform knee model on Earth, Venus, Mars and Titan. Radio Sci. **39**, RS5007 (2004). doi:10.1029/2004RS003056

T.J. Ringrose, M.C. Towner, J.C. Zarnecki, Convective vortices on Mars: a reanalysis of Viking Lander 2 meteorological data, sols 1–60. Planet. Space Sci. **163**, 78–87 (2003). doi:10.1016/S0019-1035(03)00073-3

C. Ruf, N. Renno, J. Kok, E. Bandelier, M. Sander, S. Gross, L. Skjerve, B. Cantor, Emission of non-thermal microwave radiation by a Martian dust storm. Geophys. Res. Lett. **36**, L13202 (2009). doi:10.1029/2009GL038715

Chapter 5
Jupiter and Saturn

Abstract Vigorous lightning has been observed on both Jupiter and Saturn but these gas giant planets are not expected to have global circuits, as the deep lower atmosphere is too dense for cosmic ray ionisation. Where cosmic ray ionisation occurs, charged aerosol particles are generated, making the air weakly conductive.

Jupiter and Saturn both resemble the Sun more than the Earth. Their huge atmospheres are dominated by hydrogen and helium, in which the pressure and temperature increase with depth. At pressures >40 GPa, hydrogen starts to dissociate, and becomes "metallic" and electrically conducting at pressures >300 GPa. It is probable that Jupiter's "surface" is a high-pressure liquid hydrogen "ocean" which becomes semi-conducting and then conducting as the pressure increases, with a small (Earth-sized) rocky core (Nellis 2000; Stevenson 2003). The metallic hydrogen acts as a dynamo generating the substantial Jovian planetary magnetic field, which is the highest of any body in the solar system, see Table 5.1. Despite these apparent differences from the terrestrial planets discussed above, Jupiter and Saturn have active weather systems driven by convection. Unlike the inner planets, where weather systems are driven by insolation, convection in the outer planets arises principally from internal heat generation (Atreya 1986).

Jupiter in particular has been the subject of many years of study, as its size makes it relatively easy to observe surface features from Earth. For example, Robert Hooke reported observing the motion of bands and spots on Jupiter in 1666 (Hooke 1666). Both Saturn and Jupiter have active convection and cloud layers composed of ammonia, ammonium hydrosulphide and water, resulting in their characteristic colours and banded structure (Lewis 1997; Desch et al. 2002). Clouds and convection suggest lightning activity, which has been observed on both Jupiter and Saturn. Jovian lightning was optically detected by Voyager, Galileo and New Horizons (Lanzerotti et al. 1983; Little et al. 1999; Baines et al. 2007), and Voyager observed radio signals from lightning on Saturn (Kaiser et al. 1983), with optical and radio emissions simultaneously detected by Cassini (Dyudina et al. 2010). Gas giant lightning probably originates from water clouds deep within

K. L. Aplin, *Electrifying Atmospheres: Charging, Ionisation and Lightning in the Solar System and Beyond*, SpringerBriefs in Astronomy,
DOI: 10.1007/978-94-007-6633-4_5,

Table 5.1 Comparison of planetary magnetic fields, compiled from Stevenson (2003), Ip et al. (2000) and Smoluchowski (1979)

Planet/*Moon*	Approximate surface magnetic field, Tesla (T)
Venus	$<10^{-8}$
Earth	5×10^{-5}
Mars	$10^{-9}-10^{-4}$
Jupiter	4.2×10^{-4}
Saturn	2×10^{-5}
Titan	$<10^{-7}$
Uranus	2×10^{-5}
Neptune	2×10^{-5}
Triton	Probably negligible
Pluto	Not known

the atmosphere, and is likely to be generated by similar mechanisms to terrestrial lightning (Lewis 1997; Yair et al. 1998).

From observations of the rotation period of the signals, Kaiser et al. (1983) associated "Saturn electrostatic discharges" (SEDs) with equatorial storms, and this connection was confirmed by Cassini observations. The Radio and Plasma Wave Science (RPWS) instrument has observed numerous SEDs (Gurnett et al. 2005) which were spatially and temporally correlated with southern mid-latitude storm systems seen by the Imaging Science Subsystem (Porco et al. 2005), but lightning was only detected optically on Saturn in 2010 (Dyudina et al. 2010). Whistler mode emissions were also seen by the RPWS package and have been linked to emissions from particular latitudes, but they can only be detected under specific conditions, and are not easy to associate with particular storms (Fischer et al. 2008). Saturn's long-lived and energetic storms continue to be observed by Cassini and also from Earth (Zarka et al. 2008; Yair 2012). Electrostatically-generated structures known as "spokes" in Saturn's rings have also been associated with electron beams from thunderstorms (Jones et al. 2006).

5.1 Cosmic Ray Ionisation and Ion-Aerosol Interactions

Jupiter's strong internal magnetic field screens out all except the most energetic cosmic rays, leading to a cut-off rigidity at the equator of 1 TV, whereas at Earth the cut-off rigidity is 10 GV (Bazilevskaya et al. 2008). High-energy particles of >100 MeV may influence chemical reactions in the lower atmosphere where solar UV radiation cannot penetrate (Lewis 1997). This motivated the study of Capone et al. (1979) who modified an earlier cosmic ray atmospheric ionisation model (Capone et al. 1977) to include the effect of muons, which are relatively more important at atmospheric densities >750 gcm^{-2} (~ 750 hPa). More recently, Whitten et al. (2008) modelled the electrical conductivity and charging of cloud particles in the atmosphere of Jupiter. Three cloud layers were assumed to exist from 0.1 to 6 bar, an upper layer of ammonia ice, a middle layer of ammonium hydrosulphide and a deeper layer of water ice cloud. Both pieces of work used similar approaches to calculate cosmic ray ionisation rates, finding a substantial latitudinal variation, as would be expected for a planet with a powerful internal dipolar magnetic field. The dominant terminal positive ions are expected to be $NH_4^+(NH_3)_n$ but the negative ion concentration is sensitive to the concentrations of electrophilic species, and also whether clouds are present or not (Whitten et al. 2008). In cloudy regions, free electrons could attach either to cloud particles or electrophilic species; however, it is known that no suitable electrophilic species are abundant enough (1 ppm would be needed) in the cloudy regions to compete with the cloud particles in attaching electrons. Whitten et al. (2008) carried out a sensitivity study of negative ion and electron concentrations for varying mixing ratios of electrophilic species in both the presence and absence of clouds. If clouds are not present, then even low electrophile concentrations can cause significant numbers of negative ions to be generated. Negative cluster-ions are therefore likely to be important atmospheric charge carriers above and below the cloud region. In the cloud areas, the mean charge is negative, although a bipolar distribution is expected. Conductivity within the clouds varies from 10^{-17} to 10^{-20} Sm^{-1}, with the lowest conductivity found in the deep water clouds, Fig. 5.1.

As Jovian lightning is expected to originate from water clouds, some of the findings from this "fair weather" ion-aerosol modelling could be relevant to the

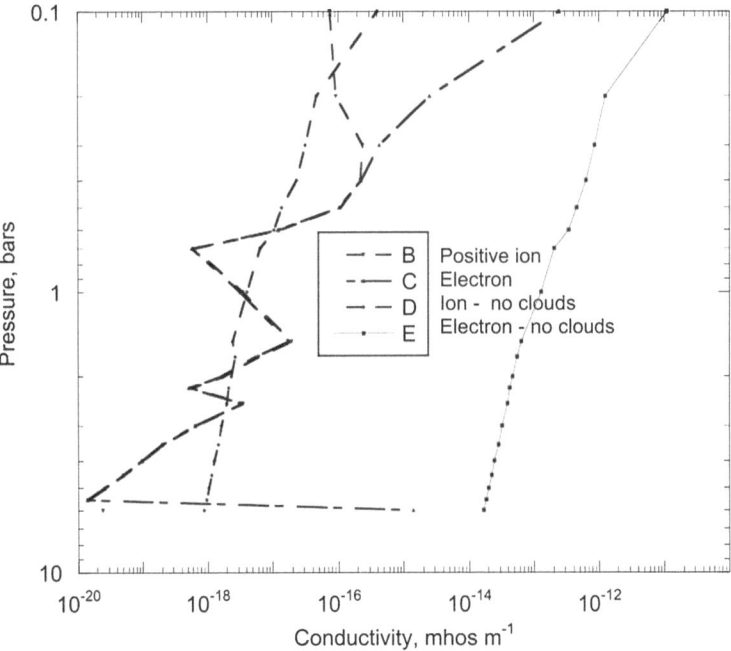

Fig. 5.1 Modelled air conductivity on Jupiter, showing the effects of clouds, after Whitten et al. (2008)

physical processes leading to lightning generation (Yair et al. 1998). The Whitten et al. (2008) study shows that unlike Venus, the calculated in-cloud conductivity appears low enough not to limit charge separation, although the assumption of thundercloud composition as monodisperse 5 μm ice cloud is clearly simplistic. Little is known about the particle size distribution so deep in the Jovian atmosphere; the modified gamma function used by Yair et al. (1998) might be a more realistic approximation.

5.1.1 Role of Ion Chemistry

Modelling the ion chemistry indicated that cosmic ray ionisation could ultimately enhance synthesis of molecules like C_3H_8 (propane) and CH_3NH_2 (methylamine/ amino methane), but solar UV ionisation would dominate above the troposphere, where the clouds are (Lewis 1997; Capone et al. 1979). Because of the large magnetic field, Jovian cosmic ray chemistry could show a strong latitudinal variation. A similar model applied to Saturn's atmosphere found that $C_2H_9^+$ dominated in the lower atmosphere, but no chemical predictions were made (Capone et al. 1977), possibly because Saturn's atmosphere has a higher relative concentration of hydrogen and a lower abundance of reactive trace species than Jupiter (Lewis 1997).

5.2 Global Atmospheric Electric Circuit

Whilst lightning, cosmic ray ionisation and mobile charge carriers exist in the atmospheres of Jupiter and Saturn, there are two reasons why the structure of a gas giant planet prevents a global electric circuit. First, cosmic ray ionisation decreases as atmospheric pressure increases, and a region of the deep atmosphere where ionising species cannot penetrate is predicted (Capone et al. 1979). Charged species could only enter this region by transport downwards, which is unlikely due to the internal heat source driving convection, and suggests the deep Jovian atmosphere may be electrically neutral. Another condition for a global circuit is that a planetary surface needs to be more conductive than its atmosphere (Aplin et al. 2008). Hydrogen is compressed into an insulating liquid before it starts to dissociate and become electrically conductive, so the Jovian "surface" is probably electrically insulating (Stevenson 2003). Sentman (1990) predicted that lightning would cause resonances at 1–2 Hz in the cavity between Jupiter's surface and ionosphere. However, subsequent laboratory work improved understanding of the behaviour of hydrogen at high pressure (Weir et al. 1996), and Sentman's (1990) assumptions about the conductivity of the Jovian interior may now be outdated. It is concluded that not all the necessary conditions for a global circuit, a conductive atmosphere, and a conductive surface relative to the atmosphere, appear to be fulfilled for Jupiter and Saturn, although all the other criteria are met. Ring-planet coupling may provide an alternative "circuit" model, but would require a mechanism for the particles in the rings to transfer charge or energy back to the planet.

References

K.L. Aplin, R.G. Harrison, M.J. Rycroft, Investigating earth's atmospheric electricity: a role model for planetary studies. Space Sci. Rev. **137**, 1–4, 11–27 (2008). doi:10.1007/s11214-008-9372-x

S.K. Atreya, *Atmospheres and ionospheres of the outer planets and their satellites* (Springer, Berlin, 1986)

K. Baines, A.A. Simon-Miller, G.S. Orton et al., Polar lightning and decadal-scale cloud variability on Jupiter. Science **318**, 226 (2007). doi:10.1126/science.1147912

G.A. Bazilevskaya, I.G. Usoskin, E.O. Flückiger et al., Cosmic ray induced ion production in the atmosphere. Space Sci. Revs. (2008). doi:10.1007/s11214-008-9339-y

L.A. Capone, R.C. Whitten, S.S. Prasad, J. Dubach, The ionospheres of Saturn, Uranus and Neptune. Ap. J. **215**, 977–983 (1977). doi:10.1086/155434

L.A. Capone, J. Dubach, R.C. Whitten, S.S. Prasad, Cosmic ray ionisation of the Jovian atmosphere. Icarus **39**, 433–449 (1979). doi:10.1016/0019-1035(79)90151-9

S.J. Desch, W.J. Borucki, C.T. Russell, A. Bar-Nun, Progress in planetary lightning. Rep. Prog. Phys. **65**, 955–997 (2002). doi:10.1088/0034-4885/65/6/202

U.A. Dyudina, A.P. Ingersoll, S.P. Ewald et al., Detection of visible lightning on Saturn. Geophys. Res. Letts. **37**, L09205 (2010)

G. Fischer, D.A. Gurnett, W.S. Kurth et al., Atmospheric electricity at Saturn. Space Sci. Revs. (2008). doi:10.1007/s11214-008-9370-z

D.A. Gurnett, W.S. Kurth, G.B. Hospodarsky, A.M. Persoon, T.F. Averkamp, A. Cecconi, A. Lecacheux, P. Zarka, P. Canu, N. Cornilleau-Wehrlin, P. Galopeau, A. Roux, C. Harvey, P. Louarn, R. Bostrom, G. Gustafsson, J.E. Wahlund, M.D. Desch, W.M. Farrell, M.L. Kaiser, K. Goetz, P.J. Kellogg, G. Fischer, H.P. Ladreiter, H. Rucker, H. Alleyne, A. Pedersen, Radio and plasma wave observations at Saturn from Cassini's approach and first orbit. Science **307**(5713), 255–1259 (2005). doi:10.1126/science.1105356

R. Hooke, Some observations lately made at London concerning the planet Jupiter. Phil. Trans. **1**(14), 245–247 (1666)

W.H. Ip, A. Kopp, L.M. Lara, R. Rodrigo, Pluto's ionospheric models and solar wind interaction. Adv. Space Res. **26**(10), 1559–1563 (2000). doi:10.1016/s0273-1177(00)00098-3

G.H. Jones et al., Formation of Saturn's ring spokes by lightning-induced electron beams. Geophys. Res. Lett. **33**, L21202 (2006). doi:10.1029/2006GL028146

J.S. Lewis, *Physics and chemistry of the solar system* (Academic Press, San Diego, 1997)

L.J. Lanzerotti, K. Rinnert, E.P. Krider, M.A. Uman, G. Dehmel, F.O. Gliem, W.I. Axford, Planetary lightning and lightning measurements on the Galileo probe to Jupiter's atmosphere, in *Proceedings in Atmospheric Electricity*, Hampton, Virginia, 1983, ed. by L.H. Ruhnke, J. Latham, A. Deepak

B. Little, C.D. Anger, A.P. Ingersoll, A.R. Vasavada, D.A. Senske, H.H. Breneman, W.J. Borucki, The Galileo SSI Team, Galileo images of lightning on Jupiter. Icarus **142**, 306–323 (1999). doi:10.1006/icar.1999.6195

M.L. Kaiser, J.E.P Connerney, M.D. Desch, Atmospheric storm explanation of saturnian electrostatic discharges. Nature **303**, 50–53 (1983). doi:10.1038/303050a0

W.J. Nellis, Metallization of fluid hydrogen at 140 GPa (1.4 Mbar): implications for Jupiter. Planet. Space Sci. **48**(7–8), 671–677 (2000). doi:10.1016/S0032-0633(00)00031-3

C.C. Porco, E. Baker, J. Barbara, K. Beurle, A. Brahic, J.A. Burns, S. Charnoz, N. Cooper, D.D. Dawson, A.D. Del Genio, T. Denk, L. Dones, U. Dyudina, M.W. Evans, B. Giese, K. Grazier, P. Helfenstein, A.P. Ingersoll, R.A. Jacobson, T.V. Johnson, A. McEwen, C.D. Murray, G. Neukum, W.M. Owen, J. Perry, T. Roatsch, J. Spitale, S. Squyres, P. Thomas, M. Tiscareno, E. Turtle, A.R. Vasavada, J. Veverka, R. Wagner, R. West, Cassini imaging science: initial results on Saturn's atmosphere. Science **307**(5713), 1243–1247 (2005). doi:10.1126/science.1107691

D. Sentman, Electrical conductivity of Jupiter's shallow interior and the formation of a resonant planetary-ionospheric cavity. Icarus **88**, 73–86 (1990). doi:10.1016/0019-1035(90)90177-B

R. Smoluchowski, Origin of the magnetic fields in the giant planets. Phys. Earth Planet. Interiors **20**(2–4), 247–254 (1979)

D.J. Stevenson, Planetary magnetic fields. Earth Planet. Sci. Lett. **208**(1–2), 1–11 (2003). doi:10.1016/S0012-821X(02)01126-3

S.T. Weir, A.C. Mitchell, W.J. Nellis, Metallization of fluid molecular hydrogen at 140 GPa (1.4 Mbar). Phys. Rev. Lett. **76**(11), 1860–1863 (1996). doi:10.1007/0-306-47086-1_4

R.C. Whitten et al., Predictions of the electrical conductivity and charging of the cloud particles in Jupiter's atmosphere. J. Geophys. Res. **113**, E04001 (2008). doi:10.1029/2007JE002975

Y. Yair, Z. Levin, S. Tzivion, Model interpretation of Jovian lightning activity and the Galileo probe results. J. Geophys. Res. **103**(D12), 14157–14166 (1998). doi:10.1029/98JD00310

Y. Yair, New results on planetary lightning. Adv. Space Res. **50**, 293–310 (2012). doi:10.1016/j.asr.2012.04.013

P. Zarka, W.M. Farrell, G. Fischer, K. Konovalenko, Ground-based and space-based observations of planetary lightning. Space Sci. Revs. (2008). doi:10.1007/s11214-008-9366-8

Chapter 6
Titan

Abstract In situ measurements of Titan's atmosphere by the Huygens probe in 2005 unveiled a fascinating world, which may not have lightning, but is a strong contender for a global electric circuit driven by electrically charged methane or ethane rain. Electrical processes also appear responsible for generating the yellow–brown haze that shrouds the planet.

The atmosphere of Saturn's largest moon, Titan, has long been of scientific interest, because it was thought to resemble conditions on Earth several billion years ago. The famous experiment illustrating the formation of amino acids by electrical discharge in a synthetic "primordial atmosphere" (Miller 1953), similar to Titan's, suggested that atmospheric electrical studies of Titan may help to understand the evolution of life on Earth. Very little was known about Titan until the Voyager 1 flyby in 1980, with another huge step forward in knowledge provided by the Huygens probe, which descended into the moon's murky atmosphere on 14 January 2005 (Lebreton et al. 2005). The parent spacecraft of Huygens, Cassini, is still adding to our understanding of Titan with its regular flybys, scheduled until the end of the mission in 2017 (at the time of writing). Titan's atmospheric composition is principally nitrogen with 6 % methane, and the temperature range permits methane to exist in three phases, much like water on Earth. This led to suggestions that there was a "hydrological" methane cycle on Titan with methane clouds and rain, subsequently confirmed by Huygens (Tokano et al. 2006). There are numerous trace organic species in Titan's atmosphere, indicating that electrical discharges may have contributed to their formation (e.g. Capone et al. 1980; Lewis 1997; Desch et al. 2002; Lorenz 2002); however, no evidence for lightning on Titan has yet been found (Yair 2012). Understanding Titan is hindered by the dense haze of organic particles ("tholins"), that obscures the surface and almost totally prevents remote sensing of the lower atmosphere. Electrification within Titan's methane clouds appeared possible based on modelling (Tokano et al. 2001), but as will be described in Sect. 6.5, observational evidence is increasingly ruling out the existence of lightning. Despite the apparent lack of thunderstorms, non-convective electrification is significant on Titan due to

K. L. Aplin, *Electrifying Atmospheres: Charging, Ionisation and Lightning in the Solar System and Beyond*, SpringerBriefs in Astronomy, DOI: 10.1007/978-94-007-6633-4_6,

the presence of free electrons, and subsequent charged particle reactions, driving much of the atmospheric chemistry and aerosol formation (e.g. Tomasko and West 2008). Additionally, the existence of both rain and free electrons points to the existence of a unique global circuit driven by charged rain, analogous to the contribution made by "shower clouds" to the terrestrial global electric circuit.

6.1 Non-Convective Electrification

The first in situ measurements of any extra-terrestrial planetary atmospheric electricity were carried out by the Pressure Wave Altimetry (PWA) experiment on the Huygens Atmospheric Structure Instrument (HASI), which carried a suite of sensors to characterise Titan's atmosphere (Fulchignoni et al. 2005). The basic atmospheric structure was as expected from modelling and earlier Voyager data, with a surface temperature of 94 K and pressure of 1,467 hPa, a tropopause located at 44 km, 115 hPa, and a stratopause at 250 km, ~0.5 hPa. This chapter summarises the current understanding of non-thunderstorm electrical processes in Titan's atmosphere, mostly but not exclusively based on data from Huygens. Firstly, ion production, which is largely modelling constrained by data, is described, secondly the unexpected finding of charged aerosol in Titan's upper atmosphere, and thirdly the Huygens measurements of air conductivity and aerosol charging in the lower atmosphere are presented.

6.2 Ion Production

Titan's upper atmosphere becomes ionised by magnetospheric electrons from Saturn (when its orbit is within Saturn's magnetosphere) and solar UV radiation, but the electrons only contribute to ionisation at heights greater than 600 km, and atmospheric density prohibits penetration of the solar UV radiation below ~40 km (Molina-Cuberos et al. 1999a). Cosmic rays are therefore the dominant source of ionisation in Titan's middle and lower atmosphere. Titan has no geomagnetic field (Table 5.1) so there is no latitudinal variation of cosmic ray ionisation. The modulation of cosmic rays by the solar wind is reduced compared to planets closer to the Sun, with little ionisation variation over the solar cycle. According to Molina-Cuberos et al. (1999b) the GCR ionisation rate is insensitive to solar cycle variations below 60 km, although Borucki and Whitten (2008) indicate that there is a factor of 2 variation over the solar cycle up to 200 km. Titan's surface is varied, with both lakes and dunes, but the chemical composition of the solid surface appears to be a mixture of hydrocarbons and nitriles covered lightly by tholins precipitated from the haze (Soderblom et al. 2007). Ion production from natural radioactivity at the surface is therefore likely to be negligible. As on Earth, most of the cosmic ray ionisation near the surface is from muons produced by a cascade of sub-atomic particles from the decay of high-energy primary cosmic rays in the upper atmosphere (O'Brien 1971; Molina-Cuberos et al. 1999b). Gronoff et al. (2009) have modelled ionisation through the whole of Titan's atmosphere, from the surface to the upper atmosphere (1,600 km), where ionisation is dominated by UV radiation. This was achieved by coupling upper and lower atmosphere ionisation models for the first time so that all sources of ionisation: UV solar flux, electron impact from Saturn and atomic oxygen ion precipitation (mesosphere upwards) and cosmic rays (stratosphere and troposphere) were considered. The cosmic ray modelling employs a full Monte-Carlo simulation of the atmospheric cascade of sub-atomic particles. This approach is more sophisticated than the analytical approximation for cosmic ray ionisation used in other planetary ionisation models (e.g. Borucki and Whitten 2008; Bazilevskaya et al. 2008).

Gronoff et al. (2009) and Borucki and Whitten (2008) were able to compare their results to the Huygens measurements; this will be discussed in the section on Huygens electrical measurements below.

6.3 Charged Aerosol

One unexpected finding from the Cassini/Huygens mission was the discovery of charged aerosol particles in the upper atmosphere, and the role that these particles play in generating Titan's ubiquitous haze. The haze is made from tholins, a complex mixture of orange or brown hydrocarbons originally generated from electrical discharges in methane/nitrogen mixtures, and subsequently found to have optical properties similar to those of Titan's atmosphere (Tomasko and West 2008). The physics and chemistry of Titan's haze layers is complex and cannot be discussed in detail here, but the role of charge will be briefly considered in the context of aerosol particle formation.

Observations from the Cassini CAPS electron spectrometer instrument revealed that large, bipolar, charged aerosol particles are abundant in the upper atmosphere (ionosphere) at $\sim 1,000$ km (Waite et al. 2007; Coates et al. 2007). Waite et al. (2007) proposed that ions created by solar ultraviolet radiation and energetic particles from Saturn's magnetosphere grow to become massive negatively charged particles present at relatively high concentrations. In one flyby, particles of 10,000 amu per electronic charge were detected. Since the CAPS instrument is being used in a non-standard mode to measure mass-to-charge ratios, particle mass information is not available, but these large particles, in a size range of 10–30 nm, are statistically highly likely to be multiply charged (Coates et al. 2007). The largest particles were present at concentrations of ~ 100 cm^{-3} at 950 km, with observations of an entire size spectrum of ions from 10 to 1,000 amu per unit charge present at concentrations from 1 to 100 cm^{-3} in the observed altitude range of 950–1,150 km. These are likely to play an important role in producing the

Fig. 6.1 Formation of tholin
aerosol particles in Titan's
atmosphere from ions.
Reproduced with permission
from Waite et al. (2007)

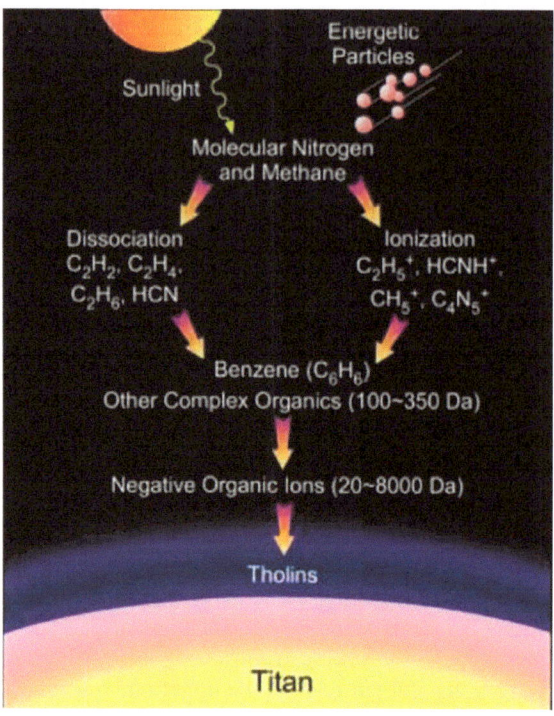

tholins that then settle gravitationally and aggregate to become Titan's many
optically active haze layers (Fig. 6.1).

6.4 Atmospheric Conductivity

The PWA experiment on Huygens consisted of two types of instrument to measure
conductivity, a relaxation probe and a mutual impedance probe. The relaxation
technique allows a conductor, charged to a fixed voltage, to decay at a rate pro-
portional to the air conductivity (e.g. Nicoll 2012). The mutual impedance probe
consists of electrodes driven by an AC current. The mutual impedance is given by
the ratio of the voltage measured at a receiving dipole to the current injected into
the medium by a nearby transmitting dipole (Hamelin et al. 2007). Although
Titan's atmospheric ionisation rates are lower than on Earth, the conductivity was
expected to be greater because of the different atmospheric composition, in particular, the presence of free electrons (Borucki et al. 1987). The relaxation probe
was therefore intended to measure the lower conductivity regions near Titan's
surface, which were expected to be similar to the terrestrial stratosphere, with
$10^{-14} < \sigma < 10^{-12}$ Sm^{-1}. In contrast, the mutual impedance probe is sensitive to
higher conductivities, 10^{-12}–10^{-7} Sm^{-1}. The relaxation probe technique also

permits derivation of the ion mobility spectrum, which is related to the ion mass spectrum (Aplin 2005; Owen et al. 2008). Two relaxation probes were used, one of which (RP2) had a substantial parallel capacitor to ensure that conductivities in the upper part of the range (in the free troposphere) were measured accurately, with RP1 intended to focus on the low conductivity regime near the surface (Grard et al. 2006). Ultimately, data were lost from RP1, and the time resolution of the mutual impedence probes was half what it should have been (Lebreton et al. 2005).

The electrical properties of Titan's lower atmosphere were expected to be sensitive to the poorly known concentration of electrophilic species. Negative ion concentration and conductivity were calculated by Borucki et al. (1987) and Molina-Cuberos et al. (2001) for different mixing ratios of electrophilic species. The maximum expected mixing ratio 10^{-11} gives similar positive and negative ion concentrations, of a few hundred cm^{-3} up to 20 km, whereas a lower mixing ratio of 10^{-15} leads to <10 negative ions cm^{-3}, with proportionally more free electrons. Borucki and Whitten (2008) subsequently suggested that the ratio should be 10^{-11} to reconcile the optical and electrical measurements made by Huygens. In terms of ion composition, Capone et al. (1980), and Borucki et al. (1987) predicted that nitrogenated cation clusters such as $NH_4^+(NH_3)_n$ and $HCNH^+(HCN)_n$ would be common. Other studies (Molina-Cuberos et al. 1999a) expected hydrocarbon ions like $CH_5^+CH_4$ to be abundant, but the 1980s work has now been shown to be more realistic (Bernard et al. 2003). The positive conductivity is related to the cluster mobility (mass), which was relatively insensitive to the detailed chemistry, and a positive air conductivity of 10^{-15} Sm^{-1} at the surface, reaching 10^{-11} Sm^{-1} at 70 km was predicted. The positive surface conductivity is similar to that on Earth, but the air conductivity increases more slowly with height than on Titan because of the shallower decrease in atmosphere density with altitude. Whilst Huygens data loss affected the lower atmosphere measurements, a peak lower atmosphere conductivity of 3×10^{-9} Sm^{-1} at the cosmic ray ionisation maximum at 60 km was recorded by both conductivity instruments, Fig. 6.2 (Hamelin et al. 2007; Molina-Cuberos et al. 2010).

The conductivity peak shown in Fig. 6.2 is related to the cosmic ray ionisation maximum, Fig. 6.3. Although the shape of the conductivity curve is qualitatively similar to predictions, the conductivity (and the related electron concentration) is a factor of 2–3 lower than expected (Hamelin et al. 2007), indicating that cluster ions could be present. Determining the cluster-ion concentration is complicated by the presence of aerosols in the lower atmosphere, which can act to either decrease the electron concentration, by attachment, or increase it, by photoionisation. For example, the Gronoff et al. (2009) model uses simpler aerosol interaction terms than Borucki and Whitten (2008), leading to a slight inconsistency in calculated electron concentrations with Huygens observations in the lower atmosphere.

Earth has a similar ionisation rate profile to Titan, with the ionisation peak at 15–26 km, (dependent on latitude and solar activity) defining a point where the atmosphere becomes sufficiently dense for most primary cosmic rays to decay and

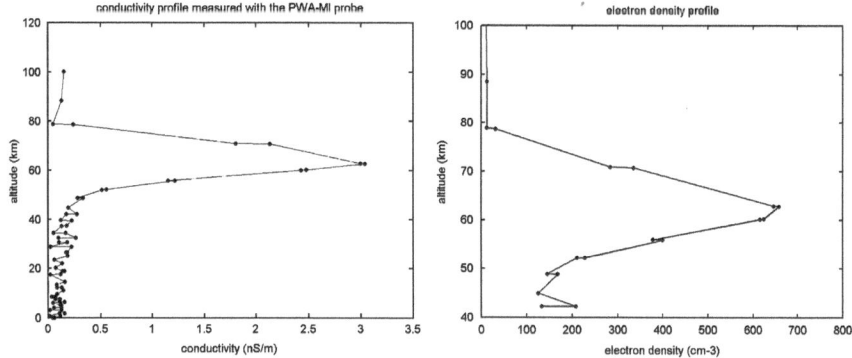

Fig. 6.2 Conductivity (*left*) and electron concentration (assuming negligible cluster-ion concentrations) (*right*) in Titan's atmosphere, measured with the Huygens Mutual Impedance Probe. Reproduced with permission from Hamelin et al. (2007). Note that the electron density is not derived at low altitudes due to lack of confidence in the conductivity data

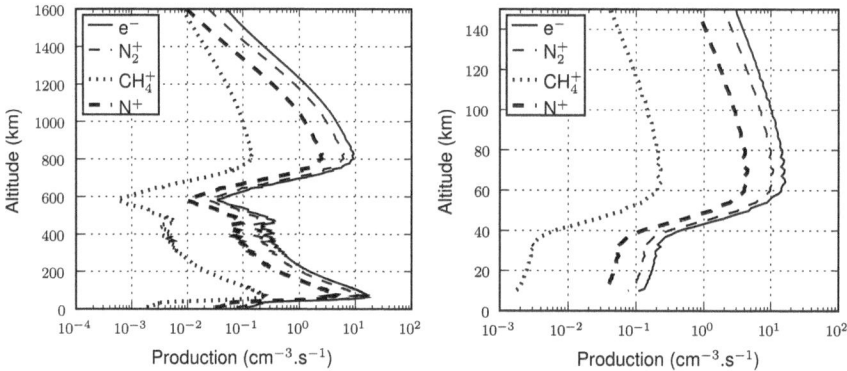

Fig. 6.3 Ionisation profiles reproduced with permission from Gronoff et al. (2009), (*left*) ionisation throughout the whole atmosphere (*right*), zoom into tropospheric ionisation by cosmic rays

produce ionising secondaries. This ionisation maximum is known as the "Pfötzer maximum" (Bazilevskaya et al. 2008), and should occur at a similar part of the Titan atmosphere to its terrestrial counterpart, i.e. the upper troposphere or lower stratosphere.[1] The Titan "Pfötzer maximum" was detected directly from Huygens conductivity measurements, whereas on Earth, it has no discernible effect on the atmospheric conductivity profile, and can only be detected in measurements of ion

---

[1] In some of the literature about Titan this peak in ionisation rate is referred to as the "ionosphere".

concentration (e.g. Rosen et al. 1982). The reasons for this are unclear. Conductivity is a function of the product of charge carrier number concentration and mobility (Eq. 2.2), where the charge carrier is an ion or electron. The most significant difference between the electrical properties of Earth and Titan is the presence of free electrons in Titan's atmosphere, so this could perhaps be relevant. However, as electrical mobility increases similarly with decreasing pressure and increasing temperature, independently of the nature of the charge carrier, this difference does not explain the propagation of the Pfötzer peak through to conductivity on Titan and not Earth.

6.5 Lightning, Schumann Resonances and the Possibility of a Global Atmospheric Electric Circuit

The modelling evidence predicting lightning, and the astrobiological significance of Titan lightning have been extensively reviewed (Griffith 2009; Yair 2012) but as there remains no evidence for lightning on Titan from numerous (72) Cassini flybys (Fischer and Gurnett 2011), it will not be discussed any further here.

An interesting 36 Hz signal spotted by the Huygens PWA instrument, consistent with the predicted frequency of Schumann resonance peaks, should also be mentioned (Fulchignoni et al. 2005). Terrestrial Schumann resonances are driven by lightning, but the consensus seems to be that Titan is the first planetary body with non-lightning driven Schumann resonances. Béghin et al. (2012) present a convincing model explaining that the 36 Hz resonance results from excitation of Titan's ionosphere by Saturn's magnetosphere. The upper boundary is assumed to be layers of charged aerosol distributed throughout the atmosphere, with the lower conducting boundary as a buried ocean.

Despite the lack of evidence for lightning, Titan could still have a global electric circuit with rain as the charge carrier (Fig. 6.4). This is analogous to the contribution of "shower clouds" (essentially charged rain), which were included in the terrestrial global circuit model first proposed by CTR Wilson. Rain has already been observed on Titan (Graves et al. 2008) and in an atmosphere containing free electrons, it is difficult to conceive of a situation where the raindrops do not become collisionally charged. This provides a charge separation mechanism, where negative charge is transferred to ground, and the cloud is left with a positive charge: the "Wilson ion capture" mechanism (e.g. MacGorman and Rust 1998). Since negative charge cannot indefinitely build up on the planet's surface, this process implies the existence of return currents elsewhere in the atmosphere, allowing for the existence of electrical discharges.

Measurements of atmospheric current flow would be required to demonstrate the existence of a Titan global circuit, as Schumann resonances are not a sufficient condition (Aplin et al. 2008; Béghin et al. 2012). On Earth, the atmospheric current flow leads to positive charge present on the top edges of layer clouds, and

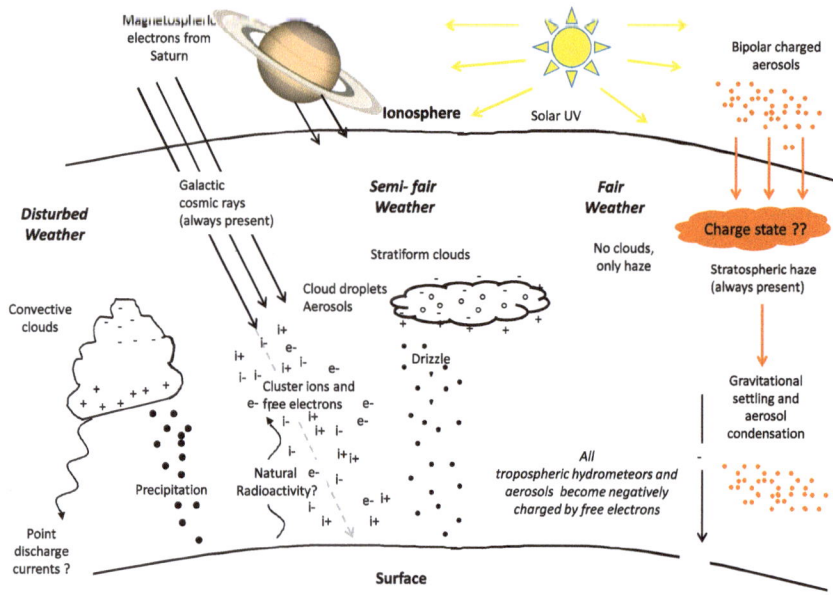

Fig. 6.4 Known or suggested processes contributing to a Titan global circuit (after Rycroft et al. 2012). Ionisation is provided mainly by solar UV radiation and Saturn's magnetosphere in the upper atmosphere, and by cosmic rays in the lower atmosphere. Regions of convective clouds, stratiform clouds and no clouds have been defined as "disturbed", "semi-fair" and "fair" weather, by analogy with Earth. All cloud droplets will become negatively charged by collision with free electrons, and precipitation, which has been observed, will leave the base of cloud with a net positive charge. The non-precipitating upper part of the cloud is assumed to retain a negative charge. Titan's haze is independent of clouds and is thought to be formed by gravitational settling of tholins formed by ion-induced upper atmospheric chemistry (Atreya, 2007). The stratospheric haze particles themselves settle to the lower atmosphere as aerosol, which will also become negatively charged. The charge state of the haze layer is unknown, but unlikely to be neutral. The flow of negative charge to the surface from precipitation and aerosol implies return currents elsewhere, perhaps through electrical discharges or other processes

negative at the bottom (Nicoll and Harrison 2010), and similar effects would be expected from a Titan global circuit. If, as is thought, rain evaporating before it reaches the ground (virga) is relatively common on Titan (e.g. Griffith et al. 2008), this could provide organised charge layers in Titan's atmosphere beyond those already associated with aerosol. The possible role of tholins and the upper atmosphere charged aerosol in any Titan global circuit still remain unknown, and unparalleled in the Solar System. The Cassini/Huygens mission may continue to identify more new atmospheric electrical processes in Titan's fascinating atmosphere.

References

K.L. Aplin, Aspirated capacitor measurements of atmospheric conductivity and ion mobility spectra. Rev. Sci. Instrum. **76**, 104501 (2005). doi:10.1063/1.2069744

K.L. Aplin, R.G. Harrison, M.J. Rycroft, Investigating Earth's atmospheric electricity: a role model for planetary studies. Space Science Reviews **137**, 1–4–11–27 (2008). doi:10.1007/s11214-008-9372-x

S.K. Atreya, Titan's organic factory. Science **316**, 843–844 (2007). doi:10.1126/science.1141869

G.A. Bazilevskaya, I.G. Usoskin, E.O. Flückiger et al., Cosmic ray induced ion production in the atmosphere. Space Sci Revs **137**, 149–173 (2008). doi:10.1007/s11214-008-9339-y

C. Béghin, O. Randriamboarison, M. Hamelin et al., Analytic theory of Titan's Schumann resonance: constraints on ionospheric conductivity and buried water ocean. Icarus **218**, 1028–1042 (2012). doi: 10.1016/j.icarus.2012.02.005

J.M. Bernard, P. Coll, A. Coustenis, F. Raulin, Experimental simulation of Titan's atmosphere: detection of ammonia and ethylene oxide Planet. Space Sci **51**(14–15), 1003–1011 (2003). doi: 10.1016/j.pss.2003.05.009

W.J. Borucki, R.C. Whitten, Influence of high abundances of aerosols on the electrical conductivity of the Titan atmosphere. Planet. Space Sci. **56**, 19–26 (2008). doi:10.1016/j.pss.2007.03.013

W.J. Borucki, Z. Levin, R.C. Whitten, R.G. Keesee, L.A. Capone, A.L. Summers, O.B. Toon, J. Dubach, Predictions of the electrical conductivity and charging of the aerosols in Titan's atmosphere. Icarus **72**, 604–622 (1987). doi:10.1016/0019-1035(87)90056-X

L.A. Capone, J. Dubach, R.C. Whitten, S.S. Prasad, K. Santhanam, Cosmic ray synthesis of organic molecules in Titan's atmosphere. Icarus **44**, 72–84 (1980). doi:10.1016/0019-1035(80)90056-1

A.J. Coates, F.J. Crary, G.R. Lewis et al., Discovery of heavy negative ions in Titan's ionosphere. Geophys. Res. Letts. **34**, L22103 (2007). doi:10.1029/2007GL030978

S.J. Desch, W.J. Borucki, C.T. Russell, A. Bar-Nun, Progress in planetary lightning. Rep. Prog. Phys. **65**, 955–997 (2002). doi: 10.1088/0034-4885/65/6/202

G. Fischer, D. Gurnett, The search for Titan lightning radio emissions. Geophys. Res. Letts. **38**, L08206 (2011). doi:10.1029/2011GL047316

M. Fulchignoni, F. Ferri, F. Angrilli et al., In situ measurements of the physical characteristics of Titan's environment. Nature **438**(8), 785–791 (2005). doi:10.1038/nature04314

R. Grard, M. Hamelin, J.J. Lopez-Moreno et al., Electric properties and related physical characteristics of the atmosphere and surface of Titan. Planet Space Sci. **54**, 1124–1136 (2006). doi:10.1016/j.pss.2006.05.036

S.D.B. Graves, C.P. McKay, C.A. Griffith, F. Ferri, M. Fulchignoni, Rain and hail can reach the surface of Titan. Planet Space Sci. **56**(3–4), 346,347 (2008). doi:10.1016/j.pss.2007.11.001

C.A. Griffith, Storms, polar deposits and the methane cycle in Titan's atmosphere. Phil. Trans. R. Soc. A **367**, 713–728 (2009). doi:10.1098/rsta.2008.0245

C.A. Griffith, C.P. McKay, F. Ferri, Titan's tropical storms in an evolving atmosphere. App. J. Letts. **687**, L41 (2008). doi:10.1086/593117

G. Gronoff, J. Lilensten, L. Desorgher, E. Flückiger, Ionisation processes in the atmosphere of Titan. Astron. Astrophys. **506**, 955–964 (2009). doi:10.1051/0004-6361/200912371

M. Hamelin, C. Béghin, R. Grard et al., Electron conductivity and density profiles derived from the mutual impedance probe measurements performed during the descent of Huygens through the atmosphere of Titan. Planet. Space Sci. **55**, 1964–1977 (2007). doi:10.1016/j.pss.2007.04.008

J.P. Lebreton, O. Witasse, C. Sollazzo, T. Blancquaert, P. Couzin, A.-M. Schipper, J.B. Jones, D. Matson, L.I. Gurvits, D.H. Atkinson, B. Kazeminejad, M. Perez-Acuyar, An overview of the descent and landing of the Huygens probe on Titan. Nature **438**(7069), 758–764 (2005). doi:10.1038/nature04347

J.S. Lewis, *Physics and Chemistry of the Solar System* (Academic Press, San Diego, 1997)

R. Lorenz, Titan's atmosphere-A review. J. Phys. IV **101**(10), 281–292 (2002). doi:10.1051/jp4:20020464

D.R. MacGorman, W.D. Rust, *The Electrical Nature of Storms* (Oxford University Press, Oxford, 1998)

S.L. Miller, A production of amino acids under possible primitive earth conditions. Science **117**, 528–529 (1953). doi:10.1126/science.117.3046.528

G.J. Molina-Cuberos, J.J. López-Moreno, R. Rodrigo, L.M. Lara, Chemistry of the galactic cosmic ray induced ionosphere of Titan. J Geophys Res **104**(E9), 21997–22024 (1999a). doi:10.1029/1998JE001001

G.J. Molina-Cuberos, J.J. López-Moreno, R. Rodrigo, L.M. Lara, K. O'Brien, Ionisation by cosmic rays of the atmosphere of Titan. Planet. Space Sci. **47**, 1347–1354 (1999b). doi:10.1016/S0032-0633(99)00056-2

G.J. Molina-Cuberos, J.J. López-Moreno, R. Rodrigo, K. Schwingenschuh, Capability of the Cassini/Huygens PWA-HASI to measure electrical conductivity in Titan. Adv. Space Res. **28**(10), 1511–1516 (2001). doi:10.1016/S0273-1177(01)00585-3

G. J. Molina-Cuberos, R. Godard, J.J. López-Moreno, M. Hamelin, R. Grard, F. Simões, K. Schwingenschuh, V.J.G. Brown, P. Falkner, F. Ferri, I. Jernej, J.M. Jerónimo, R. Rodrigo, R. Trautner, M.J. Núñez, N. Ibrahim, C. Groth, and M. Fulchignoni, A new approach for estimating Titan's electron conductivity based on data from relaxation probe sensors on the Huygens experiment. Planet Space Sci. **58**(14–15), 1945–1952 (2010). doi:10.1016/j.pss.2010.09.014

K.A. Nicoll, Measurements of atmospheric electricity aloft. Surv. Geophys. **33**, 991–1057 (2012). doi:10.1007/s10712-012-9188-9

K.A. Nicoll, R.G. Harrison, Experimental determination of layer cloud edge charging from cosmic ray ionisation. Geophys. Res. Letts. **37**, L13802 (2010). doi:10.1029/2010GL043605

K. O'Brien, Cosmic-ray propagation in the atmosphere. Il nuovo cimento **3A**(3), 521–547 (1971)

N.R. Owen, K.L. Aplin, P. Stevens, Electrical properties of ions in the atmosphere of Titan. J. Phys. Conf. Series. **142**, 012074 (2008). doi:10.1088/1742-6596/142/1/012074

I.M. Rosen, D.J. Hofman, W. Gringel et al., Results of an international workshop on atmospheric electric measurements. J Geophys Res **87**(2), 1219–1227 (1982). doi:10.1029/JC087iC02p01219

M.J. Rycroft et al., Global electric circuit coupling between the space environment and the troposphere. J. Atmos. Sol-Terr. Phys **90–91**, 198–211 (2012). doi:10.1016/j.jastp.2012.03.015

L.A. Soderblom, R.L. Kirk, J.I. Lunine et al., 'Correlations between Cassini VIMS spectra and RADAR SAR images: implications for Titan's surface composition and the character of the Huygens Probe landing site. Planet. Space Sci. **55**(13), 2025–2036 (2007). doi:10.1016/j.pss.2007.04.014

T. Tokano, C.P. McKay, F.M. Neubauer et al., Methane drizzle on Titan. Nature **442**, 432–435 (2006). doi:10.1038/nature0494

T. Tokano, F.M. Neubauer, M. Laube, C.P. McKay, Three-dimensional modelling of the tropospheric methane cycle on Titan. Icarus **153**, 130–147 (2001). doi:10.1006/icar.2001.6659

M.G. Tomasko, R.A. West, Aerosols in Titan's Atmosphere, in *Titan from Cassini-Huygens*, ed. by R.H. Brown, et al. (Springer, NY, 2008)

J.H. Waite, D.T. Young, T.E. Cravens et al., The process of Tholin formation in Titan's upper atmosphere. Science **316**, 870 (2007). doi:10.1126/science.1139727

Y. Yair, New results on planetary lightning. Adv. Space Res. **50**, 293–310 (2012). doi:10.1016/j.asr.2012.04.013

Chapter 7
Uranus and Neptune

Abstract Uranus and Neptune are both thought to have lightning and active chemistry triggered by cosmic ray ionisation. Neptune is expected to have direct condensation of supersaturated vapour onto ions, just like a "Wilson" cloud chamber. There is some evidence that Neptune's albedo is linked to the flux of galactic cosmic rays entering its atmosphere.

Uranus and Neptune are broadly similar in composition and structure to Jupiter and Saturn (Chap. 5). Their characteristic marine colours are a consequence of methane (the third most abundant species, after hydrogen and helium) absorbing the red and yellow part of the spectrum, which dominates the colours of Jupiter and Saturn. Since the Voyager 2 flybys in 1986 and 1989 respectively, new data have come from adaptive optics observations from ground-based telescopes, and from the Hubble Space Telescope. Relatively little is known about the atmospheres of Uranus and Neptune compared to the closer gas giant planets, but both have active zonal winds and changing cloud systems.

7.1 Uranus

Uranus is unique amongst the giant planets for two reasons. First, it lies almost in the plane of the ecliptic, nearly perpendicular to the other planets, so its seasons are severe with the summer hemisphere almost completely vertically illuminated by sunlight. Secondly, it has no internal heat source, so its convective activity is much lower than that of the other giants (Miner 1998). This accounts for the similar temperatures of Uranus and Neptune, even though Uranus is much closer to the Sun, and the relative calmness of the Uranian atmosphere compared to Jupiter, Saturn and Neptune. Although lightning might seem unlikely in this environment, suggestive radio emissions (Uranian electrostatic discharges, UEDs) were

K. L. Aplin, *Electrifying Atmospheres: Charging, Ionisation and Lightning in the Solar System and Beyond*, SpringerBriefs in Astronomy, DOI: 10.1007/978-94-007-6633-4_7,

observed by Voyager 2 (Zarka and Pedersen 1986). Lightning has also been suggested as a possible source for CO in the Uranian atmosphere (Encrenaz et al. 2004).

Other than hydrogen, helium and methane, the Uranian troposphere contains carbon dioxide, phosphine and stratospheric unsaturated triple bonded hydrocarbons such as acetylene (C_2H_2) and diacetylene (C_4H_2). Ethane (C_2H_6) ice clouds exist at $\sim 1,000$ hPa and there are probably hydrogen sulphide (H_2S) clouds at 3,100 hPa, possibly with water ice clouds below (Miner 1998; Encrenaz et al. 2004) The hydrocarbon chemistry is triggered by solar UV radiation resulting in haze layers, probably of solid ethane, acetylene and diacetylene at $\sim 0.05–0.13$ bar. Capone et al. (1979) predicted that the dominant ion species produced by cosmic rays in the Uranian troposphere and stratosphere were $C_3H_{11}{}^+$ and $C_2H_9{}^+$. Cosmic rays are likely to be the only ionisation at $p > 100$ hPa where solar UV radiation does not penetrate. Even though Moses et al. (1992) believed that the assumptions used in the Capone et al. (1979) model were out of date, there have been no more studies of cosmic ray ionisation in Uranus' lower atmosphere.

7.2 Neptune

Observations show that Neptune is a more convective planet than Uranus (but less so than Jupiter and Saturn) with several cloud layers, of methane, H_2S-ammonia (NH_3), water and NH_4SH (ammonium hydrosulphide), and a stratospheric haze layer (Gibbard et al. 1999). Convection is thought to transport methane ice clouds upwards to the tropopause, and there is subsidence of the stratospheric haze in other locations causing a global haze cycle (Gibbard et al. 2003). As for Uranus, the haze is composed of photochemically formed complex hydrocarbons, including higher order alkanes like propane (C_3H_8) and butane (C_4H_{10}) (Moses et al. 1992). A search for lightning by Voyager 2 revealed four possible radio emissions, and sixteen whistler events detected by two different instruments (Gibbard et al. 1999), but these were not optically corroborated, perhaps because the discharges were too deep in the atmosphere to be optically detectable (Borucki and Pham 1992). Microphysical modelling found that lightning was most likely to occur in the ammonium hydrosulphide clouds, but the study was limited by a lack of experimental data on the electrical properties of ammonium hydrosulphide, particularly the efficacy of charge transfer (Gibbard et al. 1999). More ground-based laboratory investigations could improve our understanding of the charging mechanisms.

Capone et al. (1977) calculated that the ion species formed by cosmic rays were similar to those on Uranus, dominated by $C_3H_{11}{}^+$ and $C_2H_9{}^+$. Inspired by reports of ion-induced nucleation in the terrestrial atmosphere, Moses et al. (1989) proposed a similar mechanism to explain observations of Neptune's albedo varying in antiphase with the 11-year solar cycle (Lockwood and Thompson 1986). Ions formed by cosmic rays, based on the Capone et al. (1977) model, were hypothesised to produce a methane ice haze varying with the ion production rate, but an alternative mechanism was that the colour of Neptune's aerosols could be affected by the solar UV flux (Baines and Smith 1990). After the Voyager flyby an updated and more detailed model could be produced (Moses et al. 1992) which compared the relative efficiencies of homogeneous nucleation (condensation of gas onto a nucleus of the same substance), heterogeneous nucleation and ion-induced nucleation. The low temperatures substantially limited kinetic nucleation mechanisms and increased the relative importance of electrical processes. Ion-induced nucleation was only expected to be more efficient than heterogeneous nucleation for heavy molecules with low vapour pressures, such as diacetylene (butadiyne).

Lockwood and Thompson (2002) compared the Neptune albedo with the 121.6 nm Lyman-alpha flux and the terrestrial neutron flux as proxies for the Neptune UV and cosmic ray fluxes, respectively. A statistically significant correlation was only observed between the albedo and Lyman-alpha flux, supporting the UV mechanism. The correlation between Neptune's albedo and solar activity in the photometric measurements made by Lockwood and co-workers faded after the 1990s, but Fletcher et al. (2010) reported a suggestive relationship between Neptune's mid infra-red flux density and solar activity based on four data points between 1975–2007, and Aplin and Harrison (2010) also considered Neptune's albedo and cosmic rays near Neptune measured by Voyager 2, on 1.5–3 year timescales known to be uniquely associated with cosmic rays. Enhanced variability was found in both Neptune's albedo and cosmic rays during the 1980s, when no similar signals were seen in solar UV indicators, providing support for the ion-induced hypothesis presented by Moses et al. (1992).

Further analysis would require improved cloud microphysics, cosmic ray ionisation and photochemical models. It would also be interesting to apply the Moses et al. (1992) ion-induced nucleation model to Uranus, which has similar thermal and chemical properties, to compare ion-induced nucleation with other cloud formation mechanisms.

References

K.L. Aplin, R.G. Harrison, *Ion-induced aerosol variations in the atmospheres of the outer planets*, *Proceedings Annual Aerosol Society Conference*, Southampton University, Southampton, 8–9th April 2010

K. Baines, W.J. Smith, The atmospheric structure and dynamical properties of Neptune derived from ground-based and IUE spectroscopy. Icarus **109**, 20–39 (1990)

W.J. Borucki, P.C. Pham, Optical search for lightning on Neptune. Icarus **99**, 384–389 (1992). doi:10.1016/0019-1035(92)90154-Y

L.A. Capone, R.C. Whitten, S.S. Prasad, J. Dubach, The ionospheres of Saturn, Uranus and Neptune. Ap. J. **215**, 977–983 (1977). doi:10.1086/155434

L.A. Capone, J. Dubach, R.C. Whitten, S.S. Prasad, Cosmic ray ionisation of the Jovian atmosphere. Icarus **39**, 433–449 (1979). doi:10.1016/0019-1035(79)90151-9

T. Encrenaz, E. Lellouch, P. Drossart, H. Feuchtgruber, G.S. Orton, S.K. Atreya, First detection of CO in Uranus. Ast. Astrophys. **413**, L5–L9 (2004). doi:10.1051/0004-6361:20034637

L.N. Fletcher et al., 'Neptune's atmospheric composition from AKARI infrared spectroscopy'. A&A **514**, A17 (2010). doi:10.1051/0004-6361/200913358

S.G. Gibbard, I. de Pater, H.G. Roe, S. Martin, B.A. Macintosh, C.E. Max, The altitude of Neptune cloud features from high-spatial-resolution near-infrared spectra. Icarus **166**, 359–374 (2003). doi:10.1016/j.icarus.2003.07.006

S.G. Gibbard, E.H. Levy, J.I. Lunine, I. de Pater, Lightning on Neptune, Icarus **139**, 227–234 (1999). doi:10.1006/icar.1999.6101

G.W. Lockwood, D.T. Thompson, Long-term brightness variations of Neptune and the solar cycle modulation of its albedo. Science **234**, 1543–1545 (1986). doi:10.1126/science.234.4783.1543

G.W. Lockwood, D.T. Thompson, Photometric variability of Neptune, 1972–2000. Icarus **156**, 37–51 (2002). doi:10.1006/icar.2001.6781

E.D. Miner, *Uranus: the planet, rings and satellites*, 2nd edn. (Wiley-Praxis, Chichester, 1998)

J.I. Moses, M. Allen, Y.L. Yung, Neptune's visual albedo variations over a solar cycle: a pre-Voyager look at ion-induced nucleation and cloud formation in Neptune's troposphere. Geophys. Res. Lett. **16**(12), 1489–1492 (1989). doi:10.1029/GL016i012p01489

J.I. Moses, M. Allen, Y.L. Yung, Hydrocarbon nucleation and aerosol formation in Neptune's atmosphere. Icarus **99**, 318–346 (1992). doi:10.1016/0019-1035(92)90149-2

P. Zarka, B.M. Pedersen, Radio detection of uranian lightning by Voyager 2, Nature **323**, 605–608 (1986). doi:10.1038/323605a0

Chapter 8
Triton and Pluto

Abstract Triton and Pluto both receive such low solar radiation that electrical forces are likely to be important in their atmospheres. The surprising discovery of Triton's clouds by Voyager 2 was retrospectively explained by the condensation of supersaturated nitrogen vapour onto ions, and similar processes may act at Pluto.

Triton, Neptune's largest moon and Pluto are both small bodies with similarly tenuous (surface pressures of <1 Pa) and cold atmospheres. The lack of solar energy input in the outer solar system leads to a greater role for electrical forces. For most of their orbital periods, their atmospheres are frozen, but solar heating warms their solid nitrogen surfaces (Hubbard 2003). A similar physical process forms the coma of a comet but, unlike comets, Pluto and Triton are just massive enough to retain the emitted gas as an atmosphere.

Triton is the better understood of the two bodies, as it was visited by Voyager 2 in 1989, when its atmosphere was discovered. It also yielded unexpected geological and meteorological activity with observations of geysers and clouds (Delitsky et al. 1990; Soderblom et al. 1990). Ion-induced nucleation is the only existing explanation for Triton's thin haze, predicted to form by condensation of nitrogen onto ions produced either by cosmic rays or magnetospheric particles (Delitsky et al. 1990). In Triton's cold atmosphere, larger nitrogen clusters are more stable, which encourages ion-induced nucleation once supersaturation has been reached. Delitsky et al. (1990) showed that high supersaturations of ~ 10, and associated nucleation rate increases, could be achieved by a relatively small drop in temperature because of the sensitivity of nitrogen's vapour pressure to temperature in the low-temperature regime. The coldest region, and therefore the most favourable for particle formation onto nitrogen ion clusters, is the tropopause at 9 km, where the haze is observed (Delitsky et al. 1990; Soderblom et al. 1990).

The "cometary" model of Pluto and Triton's atmosphere generation is supported by observations of rapid seasonal change. Triton's atmospheric pressure doubled from 1990 to 1998, and Pluto's atmospheric pressure has also doubled over the last 14 years (Elliot et al. 1998; Sicardy et al. 2003). The warming observed was consistent with nitrogen cycle models, although Pluto was moving

K. L. Aplin, *Electrifying Atmospheres: Charging, Ionisation and Lightning in the Solar System and Beyond*, SpringerBriefs in Astronomy,
DOI: 10.1007/978-94-007-6633-4_8,
© Springer Science + Business Media Dordrecht 2013

away from the Sun during this time. Occultation observations implied dynamical activity in Pluto's atmosphere, boundary layer effects and the presence of a morning haze layer. Elliot et al. (2003) also suggested that light extinction observed on Pluto could be related to the presence of photochemically produced aerosol particles, and this was explored further by Rannou and Durry (2009) who found that electrically neutral aerosols could readily grow by condensation and create an extinction layer. Photochemistry was not expected to contribute, but by analogy with Triton, charge could be involved via "Wilson nucleation", which would make aerosol condensation more efficient than expected. The NASA New Horizons mission was launched in 2006 and will fly past Pluto flyby about 10 years later. New Horizons has a UV spectrometer that will characterise the atmospheric temperature and pressure profiles, constrain chemical constituents and search for atmospheric haze particles (Stern 2008). It also carries an energetic particle detector. Results from these instruments should vastly increase our understanding of Pluto and its moon Charon.

References

M.L. Delitsky, R.P. Turco, M.Z. Jacobson, Nitrogen ion clusters in Triton's atmosphere. Geophys. Res. Lett. **17**(10), 1725–1728 (1990). doi:10.1029/GL017i010p01725

J.L. Elliot, H.B. Hammel, L.H. Wasserman, O.G. Franz, S.W. McDonald, M.J. Person, C.B. Olkin, E.W. Dunham, J.R. Spencer, J.A. Stansberry, M.W. Buie, J.M. Pasachoff, B.A. Babcock, T.H. McConnochie, Global warming on Triton. Nature **393**, 765–767 (1998). doi:10.1038/31651

J.L. Elliott, A. Ates, B.A. Babcock, A.S. Bosh, M.W. Buie, K.B. Clancy, E.W. Dunham, S.S. Eikenberry, D.T. Hall, S.D. Kern, S.K. Leggett, S.E. Levine, D.S. Moon, C.B. Olkin, D.J. Osip, L.C. Roberts, C.V. Salyk, S.P. Souza, R.C. Stone, B.W. Taylor, D.J. Tholen, J.E. Thomas-Osip, D.R. Ticehurst, L.H. Wassermann, The recent expansion of Pluto's atmosphere. Nature **424**(6945), 165–168 (2003). doi:10.1038/nature01762

W. Hubbard, Pluto's atmospheric surprise. Nature **424**, 137–138 (2003). doi:10.1038/424137a

P. Rannou, G. Durry, Extinction layer detected by the 2003 star occultation on Pluto. J. Geophys. Res. **114**, E11013 (2009). doi:10.1029/2009JE003383

B. Sicardy, T. Widemann, E. Lellouch, C. Veillet, J-C. Cuillandre, F. Colas, F. Roques, W. Beisker, M. Kretlow, A-M. Lagrange, E. Gendron, F. Lacombe, J. Lecacheux, C. Birnbaum, A. Fienga, C. Leyrat, A. Maury, E. Raynaud, S. Renner, M. Schultheis, K. Brooks, A. Delsanti, O.R. Hainaut, R. Gilmozzi, C. Lidman, J. Spyromillo, M. Rapaport, P. Rosenzweig, O. Naranjo, L. Porras, F. Diaz, H. Calderon, S. Carrillo, A. Carvajal, E. Recalde, L. Gaviria Cavero, C. Montalvo, D. Barria, R. Campos, R. Duffard, Levato H, Large changes in Pluto's atmosphere as revealed by recent stellar occultations. Nature **424**(6945), 168–170 (2003). doi:10.1038/nature01766

L.A. Soderblom, S.W. Kieffer, T.L. Becker, R.H. Brown, A.F. Cook, C.J. Hansen, T.V. Johnson, R.L. Kirk, E.M. Shoemaker, Triton's geyser-like plumes: discovery and basic characterization. Science **250**(4979), 410–415 (1990). doi:10.1126/science.250.4979.410

S.A. Stern, The New Horizons Pluto Kuiper Belt Mission: an overview with historical context. Space Sci. Rev. **140**, 3–21 (2008). doi:0.1007/s11214-007-9295-y

Chapter 9
Exoplanetary Atmospheric Electricity

Abstract Exoplanetary atmospheres will be slightly ionised, and some are expected to harbour lightning, just like the planets in our Solar System. The challenges for discovery of exoplanetary lightning are briefly discussed. Exoplanets are so diverse that novel roles for atmospheric electricity are expected, with a recent example given.

The study of extrasolar planets (exoplanets), those orbiting stars other than the Sun, is one of the most exciting areas of contemporary science. Since the discovery of the first exoplanets (Latham et al. 1989; Wolszczan and Frail 1992; Mayor and Queloz 1995) over 850 are now known and more are continually being discovered.[1] Earlier exoplanet detection techniques were biased towards the detection of large "hot Jupiters", massive planets in tidally locked orbits close to their host stars, and it was expected that an atmosphere would be discovered once one of these planets passed in front of (referred to here as a transit) or behind its star (occultation) (Seager and Deming 2010). This is an identical technique to that used to discover planetary atmospheres within our Solar System, as the start and end of a planetary transit or occultation causes a gradual decrease in the stellar output rather than an abrupt change, as would be expected for a planet without an atmosphere. The first detection of an exoplanet atmosphere (containing atomic sodium) was made in 2002 by Charbonneau et al. Most of what we now know about exoplanet atmospheres is based on theoretical radiative transfer calculations combined with spectroscopic observations, often from space telescopes. Whilst some exoplanets can also be directly imaged, they must be massive and distant from their host star for this to be possible. Exoplanet atmospheres were categorised by Seager and Deming (2010), on the basis of atmospheric content and presence or absence of volatiles:

---

[1] Data taken from http://exoplanet.eu/, accessed November 2012.

K. L. Aplin, *Electrifying Atmospheres: Charging, Ionisation and Lightning in the Solar System and Beyond*, SpringerBriefs in Astronomy, DOI: 10.1007/978-94-007-6633-4_9,

1. Those dominated by hydrogen and helium, similar to the Sun and other Sun-like stars, and comparable to the gas and ice giant planets in our solar system.
2. Outgassed atmospheres (i.e. those created by loss of gas from a rocky core) containing hydrogen. Some planets with masses 10–30 times greater than Earth are likely to be massive enough and cold enough to retain hydrogen, with linked chemistry (e.g. molecular hydrogen, water vapour, methane etc.). This type of planet does not exist in our solar system.
3. Outgassed atmospheres dominated by carbon dioxide. These planets can arise either from categories (1) or (2) above. Earth fits this category, but its CO_2 content was dissolved in the ocean and ended up in limestone. The composition of this sort of atmosphere depends on the interior geology.
4. "Hot super Earths" with temperatures $\sim 1,500$ K. These planets are hot enough for all volatiles to have been lost, and contain minerals such as silicates from calcium, aluminium and titanium. This type of atmosphere has no analogy in the solar system.

In this chapter the different genres of exoplanet atmosphere will be briefly summarised, with any possibilities of electrical activity, on the basis of the categories defined in Chap. 1. The discussion is biased more towards the hot Jupiter exoplanets, as their large size and deep atmospheres make them easier to observe. There is, of course, much scientific interest in the more Earth-like exoplanets, but little is known about them yet. The electrical phenomena can be categorised as those similar to processes acting on solar system planets, and those that do not have any parallels within our solar system.

9.1 Processes Analogous to Solar System Planets

All exoplanets have some cosmic ray ionisation throughout their atmospheres, which will create positive ions, electrons, and if electrophiles are present, negative ions. Ion-aerosol chemistry and physics models could in principle be applied to any exoplanet, but these studies do not yet exist, possibly because of lack of knowledge of chemical trace species.

If any particles are present, ions and electrons will attach to them creating charged aerosol. There is evidence for haze in the upper atmospheres of some hot Jupiters, first identified by Rayleigh scattering in transmission spectra (Pont et al. 2008; Sing et al. 2011). The haze is expected to consist of magnesium silicate grains of ~ 0.3–1 μm, and is likely to become charged by collisional interactions and, potentially, by triboelectrification (like Martian dust charging and terrestrial volcanic ash charging). Helling et al. (2011) have suggested that electrical discharges could be generated between dust grains in exoplanet atmospheres, and by analogy, lightning could also be expected on Jupiter-like exoplanets. However, the "hot Jupiters" are too hot for water to exist outside the vapour phase, ruling out the terrestrial thundercloud charging processes that are assumed to generate lightning on Jupiter and Saturn. The coolest exoplanets that can currently be observed are at ~ 900 K, with future instrumentation expected to resolve transmission spectra at ~ 300 K (Seager and Deming 2010). The quest for a "habitable super-Earth" is a strong scientific driver for reducing the minimum detectable exoplanet temperature, so a Jupiter-like or Earth-like exoplanet with suitable atmospheric conditions for lightning may be discovered in the next few years. Some exoplanets almost certainly have lightning, but detection of a faint signature from deep within an atmosphere is a non-trivial problem. The most promising detection technique is probably spectroscopically, from lightning-induced chemical changes, similar to the arguments made by Krasnopolsky (2006), who claimed that NO in the Venusian atmosphere could only originate from lightning. It should be in principle possible to search for spectral lines from in the hydrogen Balmer series, or the 588 nm helium line (Borucki et al. 1996) in Jupiter-like exoplanet atmospheres excited by discharges, either from water thunderclouds or mineral clouds. Direct detection of the electromagnetic signatures from lightning seems difficult, given the distance of these planets.

9.2 Processes Unique to Exoplanets

Batygin et al. (2011) present a novel electrical mechanism that may explain observations of hot Jupiters, which are often larger than expected from theory. The hot temperatures ($\sim 2,000$ K) will ionise the alkali metals like sodium known to be present (Seager and Deming 2010), making the atmosphere electrically conductive ($10^{-3} < \sigma < 1$ Sm^{-1} depending on the temperature). The planet is heated by radiant energy from its star, and the latitudinal dependence of this heating creates strong equatorial jets, moving faster than the planet's rotation, at ~ 1 kms^{-1}. Advection of conductive gases in these jets within the planetary magnetic field induces electromotive forces that lead to current loops being created between the atmosphere and deep interior. The Ohmic heating from the current loops interacts with the atmospheric dynamics, with the Lorentz force acting to damp the local circulation. Batygin et al. (2011) model the combined thermodynamic and radiative effects in an attempt to show that this effect accounts for the "inflated" radii of many hot Jupiters.

This is a new paradigm for atmospheric electricity, where electrical effects are intimately linked with dynamics in a way unknown in our solar system. Instead of having a modest effect on, for example, clouds, or atmospheric chemistry, the electrical properties of the atmosphere could modify the radius of the planet itself as it evolves. Other roles for atmospheric electricity are likely to be revealed, given the demonstrable diversity of exoplanets and the rate at which our understanding of them is developing.

References

K. Batygin, D.J. Stevenson, P.J. Bodenheimer, Evolution of ohmically heated hot Jupiters. Ap. J. **738**(1), 1–10 (2011). doi:10.1088/0004-637X/738/1/1

W.J. Borucki et al., Spectral irradiance measurements of simulated lightning in planetary atmospheres. Icarus **123**(336–344), 0162 (1996). doi:10.1006/icar.1996.0162

D. Charbonneau et al., Detection of an extrasolar planet atmosphere. Ap. J. **568**(1), 377–384 (2002). doi:10.1086/338770

C. Helling et al., Ionisation in the atmospheres of brown dwarfs and extrasolar planets I. The role of electron avalanche. Ap. J. **727**, 4 (2011). doi:10.1088/0004-637X/727/1/4

V.A. Krasnopolsky, Chemical composition of Venus atmosphere and clouds: Some unsolved problems. Planet. Space Sci. **54**, 1352–1359 (2006). doi:10.1016/j.pss.2006.04.019

D.W. Latham et al., The unseen companion of HD114762—A probable brown dwarf. Nature **339**(6219), 38–40 (1989)

M. Mayor, D. Queloz, A Jupiter-mass companion to a solar-type star. Nature **378**, 355–359 (1995). doi:10.1038/378355a0

F. Pont et al., Detection of atmospheric haze on an extrasolar planet: the 0.55–1.05 μm transmission spectrum of HD 189733b with the hubble space telescope. Month. Not. Roy. Ast. Soc. **385**(1), 109–118 (2008). doi:10.1111/j.1365-2966.2008.12852.x

S. Seager, D. Deming, Exoplanet atmospheres. Ann. Rev. Astron. Astrophys. **48**, 631–672 (2010). doi:10.1146/annurev-astro-081309-130837

D.K. Sing et al., Hubble space telescope transmission spectroscopy of the exoplanet HD 189733b: High-altitude atmospheric haze in the optical and near-ultraviolet with STIS. Month. Not. Roy. Ast. Soc. **416**(2), 1443–1455 (2011). doi:10.1111/j.1365-2966.2011.19142.x

A. Wolszczan, D.A. Frail, A planetary system around the millisecond pulsar PSR1257 + 12. Nature **355**, 145–147 (1992). doi:10.1038/355145a0

Chapter 10
Conclusions

Abstract In this chapter the planetary atmospheric electrical systems are summarised and priorities for future work are considered.

The planetary atmospheric electrical systems within our solar system fall naturally into three groups, similar but not identical to the traditional classifications of terrestrial, gas giant and outer planets. The "terrestrial" planetary atmospheric electric systems are, like the terrestrial planets, those most similar to Earth. These planets have well-defined surfaces and charged particle populations, with a high probability of electrostatic discharge. They are consequently the most likely group to fulfil the conditions for a global electric circuit like Earth's. The criteria for a global circuit defined in Chap. 1, and the evidence for each of them in the planetary atmospheres discussed are summarised in Table 10.1. The classical taxonomy includes Earth, Venus and Mars but the atmospheric electrification classification scheme must also include the satellite, Titan, which has similar atmospheric pressure and composition to Earth. A Martian global circuit has been proposed; it is comparable to the terrestrial model, but driven by dust storms with the opposite dipole structure to terrestrial thunderclouds. Data from the Cassini-Huygens mission has not found lightning on Titan, but hints at a global electric circuit based on charged rain, a fascinating variation on the model proposed by CTR Wilson.

Jupiter, Saturn, Uranus and Neptune all have lightning and active weather systems. It has been suggested that the condensation of gases onto ions, ion-induced nucleation, could be relevant, particularly in the atmosphere of Neptune, and possibly also on Uranus. Despite the existence of ions, aerosol, polar molecules and convection, it is difficult to apply the terrestrial model of a global circuit to these planets because of the probable absence of a conducting surface. A different electrical model for the giant planetary atmospheres may be appropriate.

Both the atmospheres of the small outer solar system bodies, Pluto and Neptune's satellite Triton, form and vanish seasonally. Studies of Triton benefited from a 1989 visit by Voyager 2, in which its atmosphere and thin tropospheric haze layer were detected, but studies have been complicated by a lack of

K. L. Aplin, *Electrifying Atmospheres: Charging, Ionisation and Lightning in the Solar System and Beyond*, SpringerBriefs in Astronomy,
DOI: 10.1007/978-94-007-6633-4_10,
© Springer Science + Business Media Dordrecht 2013

Table 10.1 Summary of global circuit criteria met by solar system planetary atmospheres

Planet/ Moon	Polar molecules	Charge separation	Conducting upper and lower layer?	Mobile charged particles in lower atmosphere
Venus	Yes	Possible	Yes	Yes
Earth	Yes	Yes	Yes	Yes
Mars	Yes	Probable, but not observed	Upper: yes Lower: doubtful	Yes
Jupiter	Yes	Yes	Upper: yes Lower: doubtful	Probably not in deep atmosphere
Saturn	Yes	Yes	Upper: yes Lower: doubtful	Probably not in deep atmosphere
Titan	Yes	Expected, but not observed	Upper: yes Lower: probably	Yes
Uranus	Yes	Yes	Upper: yes Lower: doubtful	Probably not in deep atmosphere
Neptune	Yes	Yes	Upper: yes Lower: doubtful	Probably not in deep atmosphere
Triton	Not detected	Probably not	Not known	Probably
Pluto	Not detected	Probably not	Not known	Probably

observational data. Pluto's atmosphere was only discovered in the 1980s from telescope occultation observations and with great ingenuity, the existence of atmospheric aerosols has recently been deduced from ground-based telescope studies, possibly as a haze layer. Triton's haze layer is thought to have been formed by ion-induced nucleation, and similar mechanisms could be possible on Pluto.

The same atmospheric electrical processes act across and beyond the solar system. Lightning has been detected on many other planets, and Venus, Mars and Titan could have quasi-terrestrial global circuits. Jupiter and Saturn's electrical systems are poorly understood and may have to await further understanding of the nature of their high-pressure interiors. Ion-induced nucleation could occur on Venus, Neptune, Triton and possibly Uranus and Pluto. Ion-mediated nucleation has been observed on Earth, but there have not yet been any attempts to predict or identify it in other planetary atmospheres. In the cold, yet rapidly evolving, atmospheres of Triton and Pluto, there are so few forces acting that electric charge may be relatively important. In the rapidly expanding catalogue of extra-solar planets, a new and critical role for atmospheric ionisation has been proposed that is unparalleled in our solar system. Additionally, exoplanets are expected to have charged aerosol and lightning, and observations are eagerly anticipated.

Clear needs for further research can be identified based on a comparative approach to the themes discussed.

- *Direct measurements*: Electrostatic measurements on Mars are urgently required to characterise an environment that may soon be visited by humans. The role of charged particles in atmospheric chemistry and meteorology could be investigated simply with minimal instrumentation—one relaxation probe—to measure atmospheric conductivity and electric fields on future planetary atmosphere missions.

- *Application of terrestrial atmospheric physics*: Though there is still a paucity of terrestrial experimental atmospheric electrification data, model predictions are more easily tested on Earth. Models validated for Earth's atmosphere, like the ones used to predict ion-mediated nucleation, could be redeveloped to enhance our understanding of other electrically similar worlds like Titan and Venus. Ground-based laboratory experiments to investigate terrestrial thundercloud charging could be extended to study the electrical properties of charge-separating species in other atmospheres.

- *Development of planetary atmosphere theoretical work*: The theoretical basis for predicting ion-induced nucleation on Neptune and Triton could be usefully extended to investigate Uranus and Pluto, respectively. Existing models could be applied to exoplanetary atmospheres once enough data is available. Modelling the cosmic ray flux in exoplanet atmospheres may present another challenge.